CHINESE TREE PEONY

图书在版编目(CIP)数据

中国牡丹品种图志:英文/王莲英主编.-北京:
中国林业出版社,1998.5
ISBN 7-5038-2019-5

I.中⋯ Ⅱ.王⋯ Ⅲ.牡丹-品种志-中国-图集-英文
Ⅳ.S685.11-64

中国版本图书馆 CIP 数据核字(98)第 09068 号

中国林业出版社出版
(100009 北京西城区刘海胡同7号)
深圳新海彩印有限公司印刷
新华书店北京发行所发行
1998年7月第1版 1998年7月 第1次印刷
开本:215mm×285mm 印张:14

Chinese Tree Peony
Wang Lianying et al., 1998
Includes reference and index.
1. Chinese⋯ 2. Wang⋯ 3. Tree peony - Varieties
 - China - Pictorial record

Copyright© 1998 by China Forestry Publishing House, 7, Liuhai Hutong, West City District, Beijing, P. R. China. All right reserved. Unless permission is granted, this material shall not be copied, reproduced, or coded for reproduction by any electrical, or chemical process, or combination thereof, now known or later developed.

CHINESE TREE PEONY

THE PEONY ASSOCIATION OF CHINA

Wang Lianying et al.

CHINA FORESTRY PUBLISHING HOUSE

Preface to the English Edition

My English translation of the *Chinese Tree Peony* was indirectly inspired by the late Professor Joseph Needham of Cambridge. When I visited him in 1983 I was deeply moved by his great contribution to the introduction of Chinese science and culture to the world in his 20 volumes of books. I since then have determined to follow his example and some day to translate Chinese scientific books of the same value. While in England I saw many beautiful plants of different origins, among which there is a great proportion of Chinese flowers. It is a great pity, however, that English literature written by Chinese experts in this field could rarely be seen even in the Chinese style greenhouse at Royal Kew Gardens, where more than 1700 plants introduced from China are growing. It had been my dream since then to make the knowledge of Chinese flowers widely available for the whole world. When I read the Chinese version of this book mainly written by Wang Lianying, I understood it would be a chance to realise my dream.

I focused my painstaking work on building a bridge for the west to understand Chinese culture in two ways. The information concerning typical Chinese background originally for Chinese readers was tried to be explicitly conveyed additionally with brief explanations. But a great part has been cancelled to fit the Chinese layout, although the English version was revised several times to retain the greatest information. The second way, which was difficult for a person specialised in English teaching methodology, was to accurately describe the local Chinese techniques in tree peony cultivation and other relevant information, because there are to my knowledge no English equivalents or translated Chinese peony books available.

Chinese tree peonies are still growing in a fairly mysterious kingdom in many fields. For instance, the ecological study of their reforming functions to a certain environment and the phytocoenological revelation of regulation of a peony cultivar occurrence and development in a plant community will benefit the theory of environment conservation. Moreover, the nature of formation, accumulation and inversion of its useful chemical elements will result in greater human interests.

The general statements were translated by myself, Ms Wu Jiangmei. The cultivar statements were translated by Mr Chen Xinlu. Mr Ran Dongya also made a great contribution. Special thanks are given to Mr Will McLewin, hellebore and peony specialist, of Phedar Research and Experimental Nursery, and Ms Helen Nicolson, English language consultant, of Herunic Services, for their diligent and scholarly revision of the English text for publication.

Wu Jiang Mei, *Language Training Centre, Beijing Forestry University*
May 25th, 1998

PREFACE

The tree peony, or Mudan, is a very beautiful plant originating in China. The choice of Mudan in 1994 as the Chinese national flower was yet another honour to be added to the accolades and honorific titles bestowed on it over the centuries. For example, the names Fu Gui Hua and Bai Liang Jin given to tree peonies indicate that the flower is the symbol of happiness and good luck. The plant symbolises the characteristic temperament and cultural features of the Chinese nation, the prosperity of the country, and the diligence of the people.

It was not until the period between the Qin and Han Dynasties (221BC – 220 AD) that Mudan – the tree peony – was differentiated from Shaoyao – the herbaceous peony. Mudan cultivation thrived from 220 to 589 AD, and has continued for more than 1 500 years. During this time, a gradual process of propagation and selection has led to the wide range of tree peony cultivars available now.

Present-day cultivars have tremendously varied characteristics resulting from the range of original species involved and from the different ecological environments. Our research work has determined four Cultivar Groups: the Central Plains, the Northwest, the Southern Yangtse and the Southwest Groups. Of these, the Central Plains Cultivar Group is the richest in number of cultivars with varied flower forms, colours and length of cultivation history.

This *Chinese Tree Peony* has been written under the general editorship of Ms. Wang Lianying, who is a Professor at Beijing Forestry University and President of the Mudan and Shaoyao Association of the China Flower Society. Thanks to the co-operative efforts of peony experts, professors and experienced growers from Beijing, Luoyang and Heze, it is an authoritative and comprehensive monograph. The book presents botanical, historical and current information on every aspect of Mudan in China. It also gives a detailed and invaluable catalogue of four hundred Central Plains Group cultivars with photographs and descriptions.

I accepted with pleasure the invitation to write this preface. I sincerely hope that the book will play an important part in promoting the development and success of Mudan in both China and the rest of the world.

Yu Heng,
Shandong Agricultural University,
August 1996

INTRODUCTION

The formal writing of the *Chinese Tree Peony* started after the tree peony – Mudan – was chosen as the national flower of the People's Republic of China. It is the culmination, however, of many years' work in tree peony production and research in China. The writing of this book has three main aims:

First, to express our thanks to people all over China for their love of the tree peony. Throughout 1994, a year of well-planned and organised activities was mounted to encourage people's interest in flowers and plants and to further understanding of the flora of China. This included a nationwide poll to decide on the national flower of the country, and the results demonstrate once again the Chinese people's longstanding love of the peony – it won the vote in 29 out of the 31 provinces, autonomous regions and municipalities.

This love of tree peonies in China is firmly rooted in history. As early as the Tang Dynasty (618-907AD), which was a period of national strength, a prosperous economy and a flourishing culture, Mudan was honoured with the titles "National Beauty and Heavenly Fragrance" and "The King of Flower". The peony was in fact given the status of national flower in Tang, and also in the literature of the Ming and Qing Dynasties. Contemporary dictionaries of Chinese list the peony as the national flower, and it also emerged as the favourite in the "Selection of Ten Nationwide Famous Flowers" competition run by the journal *Popular Flowers*. All these facts are convincing confirmation of the Chinese people's love, both past and present, for the peony.

As the national flower, the peony symbolises love of peace, pursuit of happiness, and the fight against despotic power. It can be taken to represent the nation's strengths and recent development since the introduction of the Open Door Policy. Our first aim with this book is to satisfy people's thirst for knowledge and understanding of the peony and its cultivation.

The second aim is to take stock of what we have. According to authoritative sources, both Chinese and foreign, there are about 35 peony species in *Paeonia* in the world. Within the genus Paeonia, the current literature lists five or six ancestral wild species of Mudan or tree peonies, all native to China. Botanists, taxonomists and horticulturists have been carrying out research into Mudan for the past 100 years. However, in the last ten years significant new findings have emerged, and it is high time to make additions and corrections to the existing literature. For example, there are new discoveries of wild Mudan species, bringing the number from five or six species to the present eight species and two sub-species. Their areas of distribution, environment, botanical forms, morphology, cytology and palynology have also been studied. China probably also has the largest number of horticultural cultivars of Mudan in the world. Research into their origins and classification has produced striking results, and the selection, cultivation and reproduction of new cultivars has undergone continuous development.

Our third aim is to produce a comprehensive and authoritative new record, in which new achievements and recent research have been added to earlier knowledge. We believe that only in this way can the overall perspective and the high level of Chinese research work on Mudan be accurately conveyed.

Mudan is amongst the earliest plants in the world to have been used in horticulture. In China, specialised books and records of Mudan feature among the earliest plant books. In the 800 years from 986AD, when Zhong Shu wrote the first florilegium of tree peonies in the world, to 1809, when Ji Nan finished his, about fourteen Mudan books and florilegia were published, outnumbering those on other well-known Chinese plants, in addition to general books containing sections on Mudan. After the establishment of the People's Republic of China, especially during the 1980s, local books and florilegia on Mudan were produced in Heze, Luoyang, Linxia, Tongling and other cities where peony research and production have prospered. Although some of these publications describe peony work in general, or make an in-depth study of some aspect of peonies, not one of them covers both Chinese peony history, resources, production and research, and also the scientific classification, sorting, registration and description. We, as a group of Chinese researchers, have therefore written this book to present Chinese tree peonies as a significant flower and plant as we move into the 21st century, and to commemorate the hard work and research carried out for many generations by our forefathers.

This book gives a systematic description of both wild ancestral species and many horticultural cultivars of Chinese Mudan, with their resources, distribution, classification, ecological and biological characteristics, reproduction and cultivation, and application. In this volume, 400 typical cultivars of the Central Plains Group are recorded according to recognised principles and methods, with concise accurate descriptions and photographs.

This book was written co-operatively by contributors from Heze, Luoyang and Beijing – researchers with a solid theoretical base, and producers with rich practical experience – under the guidance of the Mudan and Shaoyao (tree peony and herbaceous peony) Branch of the China Flower Society. To determine the morphological characteristics, habit and origin of each cultivar, all the writers went together to the field for observation, recording and careful checking, and consulted experienced local workers and horticulturists. During this investigation, the same cultivars with different names or different cultivars with the same name were sorted out, and some cultivars of no value were eliminated. Accuracy of representation and a systematic approach were pursued in the recording of the cultivars, and it is hoped that this book will be used as a standard and base for future determination, checking and recording of Mudan cultivars. During each stage, of individual writing, joint discussion and final revision, every effort has been made to maintain the highest standards. However, the writers are aware that there is scope for further research and experience, and are open to comments and suggestions.

The authors of the first half of the book are: Section 1 – Li Jiajue and Li Qingdao; Section 2 – Qin Kuijie; Section 3 – Qin Kuijie and Li Jiajue; Sections 4, 5 and 9 – Wang Lianying and Ran Dongya; Section 6 – Zhang Yuexian, Zhao Xiaoqing, Zhang Shuling, Lei Zengpu and Ran Dongya; Section 7 – Liu Xiang and Qin Kuijie; Section 8 – Li Qingdao and Zhang Yuexian; Section 10 – Cheng Fangyun. The cultivar descriptions in the second half are written by Zhao Xiaozhi, Zhao Xiaoqing, Li Qingdao and Liu Zheng'an.

The book was completed within the impressively short period of one year, thanks to the great efforts of all the writers, and the active support of provincial governments, directors and staff of peony nurseries and research institutes in Luoyang and Heze. The academician Dr. Wang Juyuan (our landscape predecessor) and Professor Yu Heng gave us enthusiastic encouragement and help and wrote the Preface – we give them our deep and earnest thanks.

Wang Lianying
August 1996

CONTENTS

Preface
Introduction

General Statements

Wild Species and Their Distribution and Ecological Environment 2
The Cultivar Groups and Their Distribution 8
A Brief History of Peony Cultivation 11
Varietal Classification System 15
Ecological Habitats and Biological Characteristics 21
Propagation and Cultivation 26
Peonies and Their Uses 31
Famous Peony Gardens 35
The Criteria for Recording Morphology of Cultivars 40
Introduction and Development of Chinese Peonies in Other Countries 47

Cultivar Statements

Single Form 52
Lotus Form 60
Chrysanthemum Form 76
Rose Form 104
Hundred Proliferate-Flower Form 114
Anemone Form 132
Golden-Circle Form 136
Crown Form 138
Globular Form 198
Crown Proliferate-Flower Form 202
Reference 208
Index of Mudan Cultivar Names 209

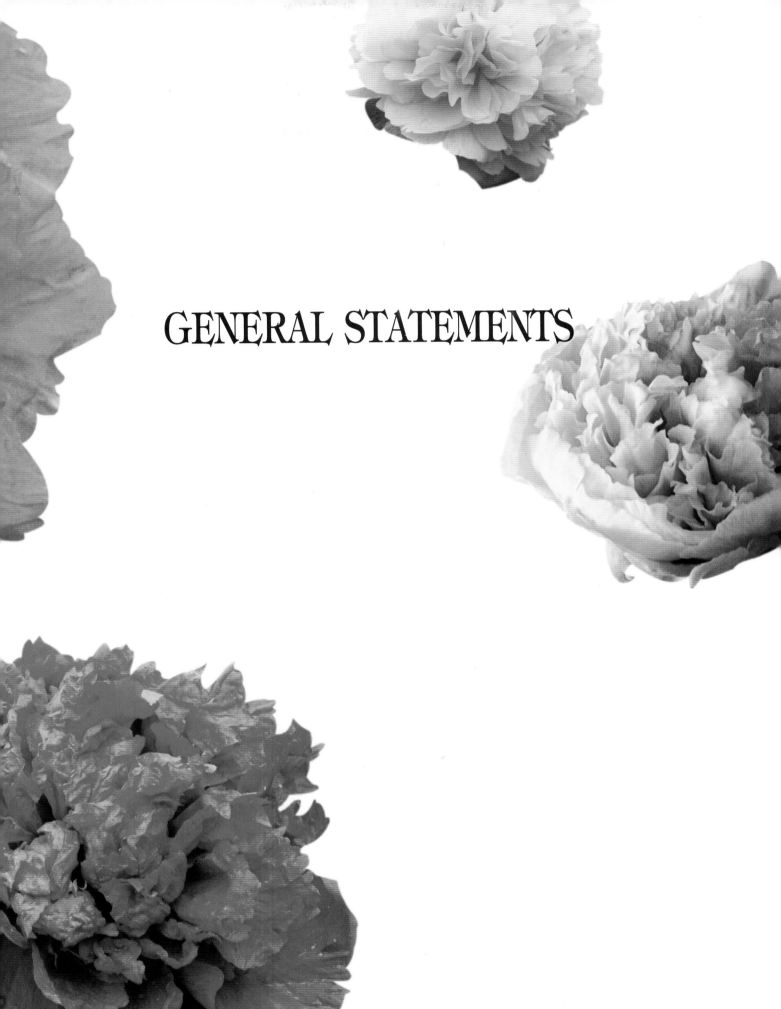
GENERAL STATEMENTS

WILD SPECIES AND THEIR DISTRIBUTION AND ECOLOGICAL ENVIRONMENT

The species that make up the Mudan, or tree peony, section of the genus *Paeonia* all originate in China. So far 8 wild species, 2 sub-species and 2 colour variants have been investigated. They can usefully be further classified into two sub-sections according to the presence or absence of a sheath that initially encloses the carpels.

Sub-section Vaginatae F. C. Stern; also called circular-leaf species group

Currently five species and one sub-species. The cup-shaped vaginate flower discs initially enclosing the carpels later split as the carpels enlarge. Ovate or ovate-lanceolate leaflets without clefts or with intermediate or deep and wide clefts. Distributed mainly in the Loess Plateau and Mountains of Qinling and Bashan, in middle and warm temperate and north sub-tropical climatic zones, the plants grow at an elevation of 1100 - 3100m. The habitats include sylvosteppe (lightly-wooded grassland), broad-leaved deciduous forests, and mixed deciduous and evergreen forests.

1. *Paeonia spontanea* (Rehder) T.Hong and W. Z. Zhao in Bull. of Bot. Res. 14(3): 238, 1994.

— *P. suffruticosa* Andr. var. *spontanea* Rehder
— *P. suffruticosa* Andr. ssp. *spontanea* (Rehder) S.G.Haw and L.A. Lauener
— *P. jishanensis* T. Hong and W.Z. Zhao.

Deciduous undershrub 1.2m high. The dried bark shows some brown colour and longitudinal veins. The annual twigs are pale green with brownish red iridescences and obscure lenticels. The two-year twigs look grey and black with fine-spotted lenticels. The five secondary leaflets on each primary leaflet of the bipinnate leaves are circular or ovate in shape with 1 - 5 clefts; no hairs on the upper surface, but a covering of tiny hairs on the under surface, which gradually come off later. The white flowers, some with faint red or light purplish-red iridescence at the base, grow singly at the top of the branches. In some stamens the filaments are dark purplish-red, but white at the top. The flower disc is also dark purplish-red. The 5 carpels have dark purplish-red stigmas. The flowering period covers the end of April and the beginning of May (Fig.1.).

The species grows mainly in south-west Shanxi and north Shaanxi. In the former, annual mean temperature in Puxian is 8.6°C, maximum 35.5°C, minimum 23.2°C, ≥10°C accumulated temperature 3002.7°C; annual mean precipitation 591.7mm, frostfree season 170 days. In Jishan and Yongji, the annual mean temperature is 13.0 to 13.8°C, ≥10°C accumulated temperature 4401.0 - 4568.3°C; annual mean precipitation 483.3 to 553.9mm; frostfree season 205 - 222 days. The main vegetation is *Quercus* in broad-leaved deciduous forests and secondary bush communities. The *Paeonia spontanea* plants grow at 1220 - 1470m on shady slopes and among bushes of open forests on shady slopes. In the mountain brown earth with pH 5.8 - 6.2, the plant communities consist of: arboreous layer (*Quercus liaotungensis*, *Q. variabilis*, *Carpinus turczaninowii*, *Tilia mongolica*, *Ulmus pumila*), bushes (*Viburnum sargentii*, *Spiraea fritschiana*, *Cotoneaster acutifolius*, *Lespedeza davurica*, *Rosa xanthina*, *Syringa oblata*, *Prunus davidiana*, *Corylus* sp.), liana (*Lonicera japonica*, *Schisandra chinensis*) and

Paeonia spontanea growing environment

Fig.1 *Paeonia spontanea*

Paeonia spontanea distribution

herbage (*Carex* sp., *Thalictrum* sp., *Adenophora* sp., *Artemisia* sp.). Around MapaoquanVillage of Jishan, the density of *P. spontanea* reaches 40 per 100m^2 (Qin Kuijie et al, 1987).

2. *Paeonia qiui* Y.L.Pei and T.Hong in Acta Phytotax. Sin. 33(1): 91, 1995

Similar to *P. spontanea*, it has regular ternate bipinnate leaves, with strictly 9 ovate secondary leaflets, purplish-red on the upper surface, usually with entire margins except for slight clefts or incisions on terminal leaflets. It grows in Songbaizhen of Sheng Nongjia Hubei at 1650m to 2010m (Fig. 2).

3. *Paeonia rockii* (S.G.Saw and L.A. Lauener) T.Hong and J. J. Li in Bull of Bot. Res. 12(3): 227, 1992.

- *P. papaveracea* Andr.
- *P. suffruticosa* Andr. var. *papaveracea* (Andr.) Kerner
- *P. suffruticosa* Andr. subsp. *rockii* S.G.Haw and L A Lauener.

This deciduous shrubby species has an erect stem 1.5m high. Leaves on the lower part of a stem are bipinnate with long petioles. The secondary leaflets are long ovate in shape on 2 - 7 primary leaflets. The bottom secondary leaflets are broadly ovate with 3 deep clefts or with none, other secondary leaflets are ovate or ovate-lanceolate. The upper surface is glabrous but soft down is sparsely scattered on the under side and densely along veins. The leaves at the top of a stem are pinnate. One flower only grows on each stem, at the apex. Ten white petals show large dark purple blotches at the base. Filaments pale yellow and flower disc yellowish-white. The 5 to 8 carpels have yellowish-white stigmas. The flowering period occurs in

Fig.2 *Paeonia qiui*

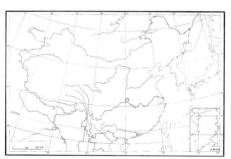

Paeonia qiui distribution

May and the seeds ripen Aug-Sept(Fig. 3).

The species is widely distributed in areas with obviously different ecological environments.

(1) The Loess Plateau Forest Area in Shaanxi and Gansu.
P. rockii grows in the Ziwuling Forest between the two provinces. This is a temperate semi-humid zone with annual mean temperature 7.4- 8.5°C, maximum 36.7°C, minimum -27.7°C; ≥10°C accumulated temperature 2600-2700°C; mean annual precipitation 500-620mm; frostfree season 110-150days. The soil is greyish-yellow and spongy or cinnamon-coloured. The plants are scattered in connected belts on northeast or shady mountain slopes under miscellaneous trees, at an elevation of 1350 to 1510m. The vegetation comprises trees and bushes including *Quercus liaotungensis*, *Populus davidiana*, *Betula platyphylla*; *Cotoneaster multiflorus*, *Lonicera pekinensis*, *Elaeagnus umbellata*, *Lonicera ferdinandii*, *Spiraea pubescens*, *Prunus tomentosa*. This species once grew in Xinglong and Maxian Mountains of Loess Plateau.

(2) Qinling and Bashan Mountainous Areas.
P. rockii has a comparatively wide distribution in the area. Since Qinling Mountain extends for 1000 kilometers, the ecological conditions vary greatly. In the western part in Gansu, *P. rockii* grows at 1100 to 2800m in broad-leaved forests or forest margins. Around Shaba In Xiaolongshan the annual mean temperature is 6.9°C, maximum 31°C, minimum -22°C; annual mean precipitation 834mm, annual mean evaporation 925.8mm; frostfree season 154 days. The soil is mainly mountain brown earth with pH 6.5.

Fig.3 *Paeonia rockii*

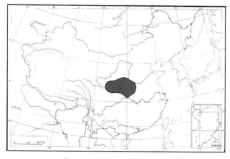

Paeonia rockii distribution

In the middle part in Shaanxi *P. rockii* is found on mountain slopes and in woods in Taibaishan Nature Conser-vation District at 1100-1800m. North or north-facing 35-degree slopes and the middle and top parts of valleys are the usual growing places, at the edges of deciduous *Quercus* forests or miscellaneous trees, or "forest windows" exposed to the sun. Cinnamon or yellow-brown earth with pH 6 - 7. Annual mean temperature 11 - 14°C, annual mean precipitation 620 - 820mm. In other places *P. rockii* grows in woods, among sparse bushes and in arid rocky cracks on sunny slopes; it is seldom seen in shady and moist gullies.

In the eastern part or in Henan, *P. rockii* is distributed at 1300-1650m around the main peak of Yangshan. The annual mean temperature in Yangshan is 12°C, maximum 43.6°C, minimum -19°C; annual mean precipitation 821.9mm; frostfree season 170 days. The mountain brown earth has pH 6.43. The average density of the plants is 6 per 100 square metres. (Zhang Yimin, Wang Jintao and Zhang Zanping 1988).

(3) Shennongjia Forest Area Located in the north west of Hubei, Shennongjia is part of the Qinling and Bashan mountainous area extending south east. *P. rockii* is distributed from 1000-2500m, growing densely at 1000-1600m in sunny open places on wooded hills. Mountain brown earth with pH 5.6-6.3, above 1500m, yellow-brown earth below 1500m. In terms of vertical distribution of vegetation, *P. rockii* is limited to mixed broad-leaved evergreen, deciduous and coniferous forest zones, with rich plant communities. Annual mean temperature 7.4 to 12.2°C; annual mean precipitation 1000-1700mm; frostfree season 153-224 days. Since *P. rocki* is distributed in such a wide area with greatly diverse ecological conditions, type differentiation in the species has taken place. Red flowers have been found in Shennongjia, pink and red have been found in Gansu and Shaanxi. Big differences in leaf structure and in form and quantity of leaflets between the north and south can be seen. In Gansu plants have 60-70 leaflets, lanceolate with complete margins. Based mainly on that leaf form, a new sub-species *P. rockii linyanshanii* (T.Hong and J.J.Li) was described in 1994 T.Hong and G.L.Osti is issued by Hongtao, 1994.The type locality is in Baokang, Hubei.

4. *Paeonia ostii* T.Hong and J.X.Zhang in Bull. of Bot. Res. 12(3): 223, 1992.

Deciduous shrub 1.5m in height. The stem bark is brownish-grey with vertical lines, the annual branches light yellowish-green. Bipinnate leaves, the primary leaflets have 5 secondary leaflets, narrow ovate-lanceolate to narrow oblong. The apex is gradually tapering while the base is wedge-shaped. Entire margins, but the terminal leaflet has 1 - 3 shallow clefts. On the upper surface thick hair grows along the central vein at the leaf base but there is no hair on the under side. Each stem bears a single flower at the apex with 11 petals, white or with a faint purplish-red suffusion. The stamens have dark purplish-red filaments and flower discs; the 5 carpels have dark purplish-red stigmas. Normal flowering period is the last three weeks of April (Fig. 4).

The species is distributed from Qinling mountain range westward to Gansu, southward to Hubei and Hunan and eastward to Anhui through Henan. The species was first discovered and introduced at Zheng Zhou in Henan. There, the species grows generally at about 1200m below open *Quercus* trees or on slopes in clusters of bushes. In some areas the distribution of *P. ostii* overlaps that of *P. rockii*, but with each growing in suitable ecological conditions. An old peony growing on a cliff on Yinping Mountain, Anhui, is believed to be *P.ostii*. *P. ostii* is the ancestor of the Fengdan group of peonies, cultivated for medicinal purposes. Since Fengdan peonies have been introduced in many places, great care is needed in the determination of the wild distribution zone of *P. ostii*.

5.*Paeonia decomposita* Hand.-Mazz.
– *P. szechuanica* Fang in Acta

Paeonia rockii

Phytotax Sin. 7(4): 302, 1958.

Glabrous, deciduous shrub 1 - 1.5m in height. The stem bark is greyish-black, annual twigs purplish-red; 2- or more-year-old branches are greyish-white but peeled off in places. Complex generally tripinnate leaves; the terminal leaflet ovate or obovate with deep clefts sometimes even reaching the base. Secondary lateral leaflets ovate or rhombus-ovate with 3 clefts or with none or with wide incisions and trilobate. One flower at the stem apex with 9 - 12 light purple to pink petals. Filaments and discs are both white, 4 - 6 glabrous carpels. The flowering period is from the last days of April to the first days of June (Fig. 5).

P. decomposita has a limited distribution, in Maerkang and Jinchuan Counties, Sichuan, at 2600-3100m. Also recorded in recent years in Wen County of Gansu.

In Maerkang, there is a clear division between dry and wet periods but without four obvious seasons. Annual mean temperature 8.6°C, mean temperature in January -1°C, minimum -17.5°C, mean temperature in July 16.5°C, maximum 34.3°C. Annual mean precipitation 753mm; annual mean duration of sunshine 2195 hours; frostfree period 120 days. *P. decomposita* grows among thorn bushes in dry valleys on south-east facing slopes. The mountain yellow earth has pH about 6.4. The main accompanying species are *Prunus davidiana*, *Rosa willmottiae*, *Rhamnus* sp., *Abelia dielsii*, *Cotoneaster soongoricus*, *Spiraea* sp., *Lonicera ferdinandii*, *Berberis polyantha* and *Syringa sweginzowii*, while the herbage consists of *Artemisia* and *pteridophytes*. In some locations the density of plants reaches 20 to 22 per 100 m².

Sub-section Delavayanae F. C. Stern; also called dissected-leaf species group

Currently 3 species, 2 colour variants and 1 sub-species. The succulent disc embraces the carpel base. Leaflets are narrow with deep clefts or incisions. Distributed in mountains at 2800-3700m in Yunnan, southwest Sichuan and southeast Tibet, the south temperate and sub-tropical climatic zones. The vegetation types there are moist broad-leaved evergreen forests and sub-alpine coniferous evergreen forests.

6. *Paeonia potaninii* Kom. in Not. Syst. Herb. Hort. Bot. Petrop., II. 7, 1921.

– *P. delavayi* Franch. var. *angustiloba* Rehd. and Wils.

Shrub 1 - 1.5m in height. Stems glabrous, light green or greyish-green. Leaves opposite or nearly opposite, biternate pinnate. Leaflets glabrous with 3-5 deep and narrow lanceolate clefts, upper surface green, underneath light green. The base of the terminal and lateral leaflets extend downwards with wing-shaped petioles. Flower red or reddish-purple (Fig. 6).

This species apparently is closely related to *P. delavayi* and *P. lutea*. Distributed among bushes on mountain slopes at 2800-3700m in western

Fig.4 *Paeonia ostii*

Paeonia ostii distribution

Fig.5 *Paeonia dcomposita*

Paeonia dcomposita distribution

Paeonia ostii

Paeonia dcomposita growing environment

Sichuan, also at 2300-2800m in Yunnan. In Yajiang County *P. Potaninii* is generally distributed in fertile black earth in river valleys and forest fringes with accompanying plants *Malus spectabilis, Broussonetia papyrifera, Cotoneaster cylindricum, Sorbaria sorbifolia, Lespedeza* sp. and *Rosa hugonis*.

(1) Variant: *P. potaninii* Komarov var. *trollioides* Stapf ex F.C.Stern.
Found in forests at 2600 - 2900m in Sichuan, 0.7m - 1.3m high; fragrant golden flowers sometimes with red interior petals. Also found in mountains in Yunnan at 1950 - 2500m.

(2) Variant: *P. potaninii* Komarov forma *alba* (Bean) F.C.Stern, with white flowers. It grows in Yunnan.

7. *Paeonia delavayi* Franch. in Bull. Boc. Bot. Fr., XXXIII. 382, 1886.

Glabrous shrub, 1.5m in height. The annual twigs are herbaceous, dark purplish-red. The leaves are ternate bipinnate, each secondary leaflet is ovate or broadly ovate, greyish-white on the under side with deep clefts forming lanceolate or oblong-lanceolate lobes. 2-5 red or purplish-red flowers with 9-12 petals, at the apex of the stem or in leaf axils. Filament dark purple. A fleshy disc embraces the base of 2-5 glabrous carpels; purple stigmas. The flowering period is in May and seeds ripen July to August(Fig. 7).

P. delavayi is found in clusters of miscellaneous trees or bushes on grassy sunny mountain slopes at 2300-3700m in Yunnan, Sichuan and Tibet. Yulong Snowy Mountain Natural Conservation District in Yunnan is chosen as an example to describe its ecological conditions. The weather is influenced by the south-west monsoon. The massif represented by Lijiang Basin (elevation 2416m) is classified as southern temperate climate. There is a clear division between wet and dry periods, without high temperatures and with a short hot period in summer, sufficient photostage in winter and spring, a long winter (135 days) and spring succeeding to autumn (230 days) with no summer. The climate from river valleys to mountain tops differs dramatically. The annual mean temperature in Lijiang is 12.6 C, at 2750m 9.7°C, maximum 26.8°C, minimum -9.6°C; the annual mean precipitation 1708mm. The corres-ponding data for the elevation of 3240m are 5.5°C, 18.8°C, -11.8°C, 1588mm. At the elevation of 2400-3000m there is mountain red earth covered by 20-30cm of surface humus in forests. The plant community consists of an arboreal layer, *Pinus densata, Pinus armandii, Picea likiang-ensis* and *Betula* sp.; shrubs, *Rhododendron yunnanensis, Quercus monimotricha, Acanthopanax senti-cosus, Lonicera* sp., *Cotoneaster acutifolius, Coriaria sinica, Viburnum cylindricum, Berberis jamesiana* var. *leucocarpa* and *Viburnum corylifolium*; and herbaceous plants, *Pteridium revolutum, Dipsacus asper* and *Oxalis griffithii*.

8. *Paeonia lutea* Detea and Franch. in Bull. Soc. Bot. XXXIII. 382, 1886.
– *P. delavayi* Franch. var. *lutea* Finet and Gagne.

Deciduous under 1.0-1.5m in height. Leaves greyish, glabrous and papery; green annually-produced twigs. The bark of 2- or more-year-old twigs peels off in places. Alternate, biternate leaves, pinnately cut, each leaflet lanceolate with entire or cleft margins. 2-5 flowers at the apex of stems, and flowers frequently produced in the 2 -3 top leaf axils. Petals yellow, sometimes with a red margin or red to purple blotches at the base. Filaments light yellow and usually three carpels. The flowering period is from the end of April to the middle of May (the middle of May in Dali, Cangshan); seed ripens September to October(Fig. 8).

Paeonia delavayi growing environment

Widespread in the central, north-western and south-western parts of Yunnan, in mountainous forest fringes at 2500-3500m. Also distributed in south-east Tibet and southwest Sichuan. *P. lutea* grows in the southwest monsoon region with clearly divided dry and wet seasons, cool summers, cold winters and great precipitation and moisture. On limestone and basalt mountains in moist broadleaved evergreen forests of *Lithocarpus variolosus* and forests of *Pinus yunnanensis*. Mountain red mountain yellearth, but in areas with good conditions, there is slightly acid yellow soil.

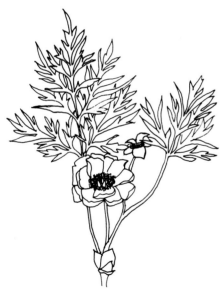

Fig. 6 *Paeonia potaninii*

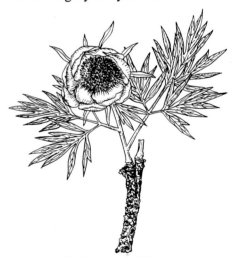

Fig. 7 *Paeonia delavayi*

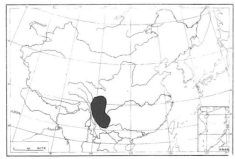

Paeonia potaninii distribution

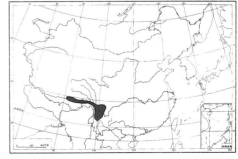

Paeonia delavayi distribution

In Kunming the annual mean temperature is 14.7°C, maximum 31.5°C, minimum -5.4°C; ≥10°C accumulated temperature 4490°C annual mean precipitation 1007mm. No fierce heat in summer or severe cold in winter, and spring lasts all year around. Rain and hot weather come simultaneously, so that the climate has the features of plateau, monsoon and subtropical zones. On shady slopes at about 2285m in the Xishan Mountain *P. Lutea* always grows in rock crevices. Many seeds are produced and seedlings can be seen everywhere. In a sample square of $50m^2$ over 100 plants grew. It lives lamellarly in mountainous shrubs in Huadianba of Cangshan, and Dali, Yunnan, with accompanying plants such as *Lithocarpus variolosus*, *Hypericum monogynum* and alpine *salix* sp., in soil with pH 6.5. Plants reach 1.9m high with 2m crown span. The dry and wet seasons are clearly divided and the vertical climatic differences are obvious with the variation of elevation.

On sunny limestone mountain slopes at an elevation of 3200m in Niya of Zhongdian County the extremely thin

Large glabrous shrub 2m in height, and able to reach 2.4m in cultivation. Annual twigs papery, yellowish-green, occasionally with red blotches. Leaves biternate, deeply toothed, yellowish-green. Branches bear 2-5 large yellow flowers with 5-8 petals and 1-3 glabrous carpels. The flowering period is April and seeds, large, black and oblate, ripen in August(Fig. 9).

Originally found in Tsanpo Gorges of south-east Tibet, also distributed on slopes at 2700-3300m in Linzhi and Milin. In the former the subalpine evergreen coniferous forests mainly contain *Picea likiangensis* var. *balfouriana* and *Abies spectabilis*, and herbage such as *Iris tectorum*, *Taraxacum mongolicum* and *Ranunculus japonicus*. The quality of the mountainous

Paeonia lutea growing environment

podzolic soil and brown earth ranges from sandy loam to light-textured soil with pH 4.9-5.7 and high organic content. In Milin brulee *Paeonia lutea* is able to form the dominant species. In Linzhi, annual mean temperature is 9.3°C, minimum 0.4°C, maximum 15.8°C°; annual mean precipitation 650mm; frostfree period 180 days.

Fig. 9 *Paeonia lutea* var. *Ludlowii*

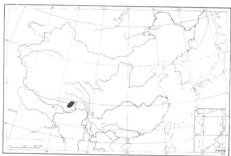

Paeonia lutea var. *ludlowii* distribution

Fig. 8 *Paeonia lutea*

Paeonia lutea distribution

surface layer of red earth is always stony. In such adverse natural circumstances *P.lutea* grows mixed with *Quercus semicarpifolia*.

Paeonia lutea Delavay and Franch. var. *ludlowii* Stern and Taylor in Bot. Mag. n.s.t. 209, 1953

Paeonia lutea var. *ludlowii* growing Environment

The Cultivar Groups And Their Distribution

In temperate, sub-tropical and tropical zones of China, different combinations of natural elements, such as topography and elevation, have provided wild and cultivated peonies with a good growing and reproductive environment and extraordinary benefits for their development. The cultivar system of Chinese peonies has been formed by many centuries of cultivation and selection. Its features are huge flowers, intense fragrance, splendid colours and great variety. Chinese peonies take first place among the flowers of the world in terms of their properties, range of flower forms, ancestral species, degree of development, adaptability, and also cultivation history. The cultivars are classified into four groups(fig.10).

Central Plains Cultivar Group

1. Historical Evolution
This group has the longest history of cultivation. The cultivation centre has always been in the Central Plains, or the middle and lower reaches of the Yellow River, ie Luoyang (during the Sui Dynasty); Xi'an (Tang); Bozhou and Heze (Ming); Heze (Qing); and latterly Heze and Luoyang. This developmental course is the main outline of the formation of the cultivar system. It shows that the Central Plains was the origin of the cultivated varieties.

2. Areas of Cultivation
The plants that comprise this group are mainly distributed in the middle and lower reaches of the Yellow River, centred on Heze, Luoyang and Beijing. They have also been planted further south between Shanghai and Hangzhou, eastwards in Qingdao and Yantai, and northwards in the open places south of the Great Wall. With suitable protection and other over-wintering techniques, they can even live and bloom normally in Heilongjiang, Gansu and Qinghai. These varieties show strong environmental adaptability and they have been planted in many foreign countries.

3. Wild Protospecies
P. spontanea and *P. rockii* are generally believed to be the wild protospecies in the formation of the Central Plains cultivar group. In recent years *P. ostii* has also been confirmed as an additional influence. The distribution in the wild of these three species is very similar to the distribution of the Central Plains cultivars. This has been beneficial for the introduction of the wild protospecies into cultivar development.

4. Main Characteristics
(1) The plants in this group are of different heights, but are generally shorter than plants in other groups. On the other hand, some *P. ostii* descendants grow tall. Their leaf shapes also vary greatly.
(2) Some varieties show a dark blotch or an iridescence at the base of petals indicating that they inherit genes from *P. rockii*, because only in *P. rockii* can large dark purple blotches be seen. *P. spontanea* and *P. ostii* show only some light purple colour at the petal base. Of the 107 traditional Central Plains cultivars, 34 have dark blotches, and another 38 have a dark iridescence. Therefore over half of the group is likely to share *P. rockii* in its origins.
(3) The Central Plains group includes nearly all the categories of flower type, which indicates a high degree of cultivar evolution.
(4) There is a wide range of colours, and both single and multiple coloured cultivars. This group surpasses the other three in the richness and fine gradation of colours.
(5) It now consists of over 500 varieties, the most among the four groups.

Northwest Cultivar Group

1. Historical Evolution
The peony in the Taoist Temple Jintian is said to be descended from a Tang Dynasty plant. If it is true, the cultivation and appreciation of the North-west peony started during Tang. When Wang Yan, the latter king of the first Shu State (919 - 925 AD) was on the throne, his maternal uncle Xu Yanqiong introduced to Chengdu an old peony plant from a temple in Qinzhou (present-day Tianshui in Gansu). Thus, peonies were grown in Tianshui very early. The cultivation of peonies in the northwest increasingly flourished during Ming and Qing. Linxia, the then centre of cultivation, was called "the Second Luoyang".

2. Areas of Cultivation
This cultivar group is distributed over most parts of Gansu, but is generally concen-trated at the middle and upper reaches of River Wei, in Longxi and Jingning. It is also distributed in Qinghai, Shaanxi and Ningxia.

3. Wild Protospecies
The Loess Plateau and the Yellow River Valley were the earliest inhabited areas of China. They are also distribution regions of *P. rockii* and *P. spontanea*. During many dynasties of cultivation, hybridisation and selection, new varieties have continuously been produced. It is believed that the main wild protospecies of the North-west peony cultivar group is *P. rockii* with some genes of *P. spontanea*. The main properties of the Northwest group

support this view.

4. Main Characteristics
(1) Plants are generally tall with annual growth of 5-15cm. There is less variation in leaf shapes, and usually long petioles.
(2) Petals have dark-purple or purple-red basal blotches and the flowers have a rich fragrance.
(3) There is a lower degree of flower form evolution than in the Central Plains group. Only 5% of the total varieties of the North-west cultivar group are proliferate-flower form. In the Central Plains population, 30% are of this form.
(4) There are approximately 100 varieties in this group. From east to west, flowering varies from late April to early June.
(5) They are particularly resistant to cold, drought and alkalinity. They grow strongly and are less troubled by disease, insects and pests. They are suitable plants for cold and dry places of the vast Northwest area. The cultivars of this group are collectively referred to as Gansu Mudan.

Southern Yangtse Cultivar Group

1. Historical Evolution
Several dozen excellent varieties were produced early in Song. The transition of centres of cultivation in each dynasty is: Hangzhou (before Song); Jiyang (Ming); Tongling, Ningguo, Hangzhou, Shanghai (Qing); Tongling, Ningguo, Hangzhou and Shanghai (present).

2. Areas of Cultivation
This group is mainly distributed in Anhui, Jiangsu and Zhejiang but centred in Tongling, Ningguo, Hangzhou, Shanghai and other towns.

3. Wild Protospecies
The main ancestor is *P. ostii*, which can still be seen along the Yangtse River. It also grows on Yinping Mountain, Anhui. The cultivars in this group are Central Plains varieties that could adapt after they were transplanted there, and also hybrids between them and *P. ostii*.

4. Main characteristics
(1) The plants are tall, with long annual twigs, and strongly branched. Xie Zaihang of the Ming Dynasty said, "There are no tall varieties in the north: the length was only three *Chi* (1m), while I saw in Jiaxing and Wujiang some over one *Zhang* (3.3m) with three or five hundred flowers on each." Thus mature plants are a magnifi- cent sight.

(2) They flower early, usually around the middle of April.
(3) There were a great number of cultivars in the past. Ji Nan in his 'Register of Peonies' (1809 AD) collected 103 varieties in Jiaxing that were known far and wide. In contrast, 20 to 30 remain nowadays.
(4) They have shallow root systems but are tolerant of humidity and heat. Because of this they have become popular for cultivation in much of southern China.

Southwest Cultivar Group

1. Historical Evolution
This group has been cultivated in Tianpeng (present Pengzhou, Sichuan) from the Tang. Many were planted in peony gardens on Danjingshan Mountain. When Luoyang varieties were introduced during the Song dynasty, the cultivation became widespread. Lu You, in 'Florilegium of Tianpeng Peonies' (1178) noted 100 varieties, of which 40 were well-known. In the book, 34 cultivars specially grown in Tianpeng were described, another 31 were mentioned only by name. Although peony cultivation declined in the Qing, it is now flourishing again.

2. Areas of Cultivation
This group is distributed in Sichuan, Yunnan, Guizhou and Tibet. The main centres of cultivation are Pengzhou and Chengdu but there are other important places. Since there are great climatic differences between the river basins and the high plateau, the characteristics of the cultivars are also varied. When sufficient data is available, it will be possible to sub-divide the group into two categories.

3. Wild Protospecies
This needs further investigation. Among the known species with rounded leaves, only *P. dcomposita* is distributed in the Southwest, but there are no reports on its use in horticulture. However, the wild distribution of *P. ostii* is adjacent to Northeastern Sichuan, and as the South-west plants are tall with long flower stalks, further study is needed to determine the possible influence of *P. ostii*. The other wild

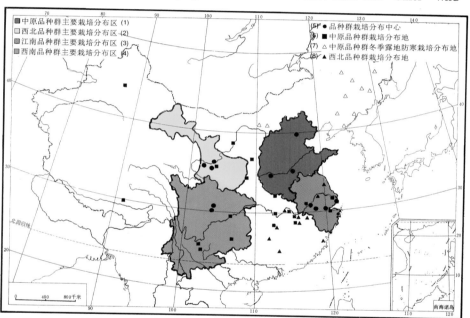

Fig. 10 The Cultivar groups distribution   (1) Central plains cultivar group distribution; (2) Northwest cultivar group distribution; (3) Southern Yangtse cultivar group distribution; (4) Southwest cultivar group distribution; (5) The centres of cultivar group distribution; (6) The distribution area of central plains cultivar; (7) The protection over-wintering distribution area of central plains cultivar; (8) The main cultivated area of northwest cultivar group.

protospecies are those with cleft leaves. From analysis of the cultivar leaf shapes, the protospecies are not likely to be those of the Southwest. In addition, there is no data on the use in horticulture of the Southwest protospecies. So although local protospecies are likely to have been involved, Tianpeng cultivars appear to be the result of breeding and selection of Central Plains and Northwest varieties that adapted to the local environment. So the wild protospecies of the Tianpeng cultivars, as in other groups, are likely to be mainly *P. rockii* and *P. spontanea*, probably with some genetic influence from *P. ostii*.

4. Main Characteristics
(1) The plants are tall and sturdy but sparsely branched and with large leaves. They have long flower stalks and purple blotches or red iridescence at the base of petals.
(2) Root systems are shallow but the plants grow well.
(3) The degree of flower form evolution is high; many varieties are of proliferate-flower form.
(4) The cultivars are generally less floriferous, but flowers are large and long-lasting. The group comprises more than ten varieties.

Summary
A. From the point of view of historical development, the Chinese peony cultivar system has formed from the evolution of one main line, the Central Plains cultivar group, and several subsidiary lines. Some peony growing areas are exceptional, for example Taiwan which has mainly Japanese varieties.
while in Yan'an, the population of those from the Central Plains.
B. To determine the genetic relationships between cultivars and their wild protospecies, further testing is needed using modern scientific techniques including those of cytology, biochemistry and palynology.
C. Although the Chinese peony system consists of four population groups which historically have developed independently from each other, the groups have never been completely isolated. Indeed, it has been common since ancient times that varieties in each group have frequently been exchangedand blended. The cultivar groups, however, maintain their own characteristics because of the differences between the local environments and the wild protospecies mainly involved. Indications of blending are most obvious in the locations where population groups are adjacent to each other.

A Brief History Of Peony Cultivation

The earliest philological record about peonies was found in 1972 at a tomb of the early Eastern Han Dynasty in an archaeo-logical excavation at Cypress County, Wuwei, Gansu Province. The medical book, carved on bamboo pieces, prescribed peony as a remedy for blood stasis. Another medical book 'Shen Nong's Herbal Classic' in the Three Kingdoms (220-265AD) is a compilation of medical experience since the two Han Dynasties. These data show that peonies had come to play a part in human life approximately 2 000 years ago through medicine. Xie Lingyun in Song (420-479AD, one of the southern Dynasties) wrote, "The peony tends to grow in waters or bamboo clusters in Yongjia" (*Collection of Xie Kangle's Essays*). Xie's comments might imply that the wild peony flower had become an object of people's interest and appreciation. Liu Yuxi of Tang said, "Yang Zihua in the Northern Qi Dynasty (550-577AD, one of the north Dynasties) definitely painted peonies" (*Record of Interesting Words From Liu Binke* by Wei Xuan of Tang Dynasty). They show that the peony had been gradually introduced to cultivation as an ornamental plant for about 1 500 years. However, in "Painting of Biographies of Outstanding Women" by Gu Hutou (about 344-405AD) from Yu Renzhong's edition in Song, *Tree Shaoyao* was drawn as a cultivated garden plant. Gu Hutou (or Gu Kaizhi) was a great painter in the Eastern Jin Dynasty (317-420AD). Thus, pictorial evidence shows that peonies were cultivated at least 100 years earlier.

Mudan (tree peony) and Shaoyao (herbaceous peony) are placed in the same genus, Paeonia, in the family Paeoniaceae. The leaves and flowers are similar, but the former is an arboreal plant with a permanent woody branch structure, while the latter is a herbaceous plant with new growth each year from ground level. Before Qin, there was no clear distinction between mudan and shaoyao. "Based on Shaoyao, a tree peony was initially called tree shaoyao" (*Complete Annals* by Zheng Qiao, the Song). According to the *Summary of Complete Annals*, the cultivation of Shaoyao can be traced to the periods of Xia, Shang and Zhou.

The earliest records of varieties of ornamental peonies are from the Sui Dynasty. When the Emperor Suiyang was on the throne (605-618AD), he "cleared land about 100km in circumference, for the Garden Xiyuan. In his Imperial Edict, all kinds of animals and plants in the territory should be carried to the capital (present Luoyang). Twenty boxes of peonies were paid as tributes from Yizhou (present County Yi, Hebei). In Chang'an (present Xi'an) the capital of Tang, peonies were gradually planted. As precious plants, they could be cultivated only in imperial gardens. Shu Yuanyu of Tang in his *Preface to Ode to Peonies* states, "Empress Wu Zetian ordered peonies to be transplanted to the Shangyuan Garden of the Palaces. From then on, peonies flourished in the capital." The poet Li Bai composed three poems to the given tune of *Qing Ping Yue*, extolling the charming shapes and gorgeous colours of peonies. His poetry has won universal praise as a masterpiece for more than a thousand years. Since then, peonies have enjoyed a considerable reputation. Furthermore, in 825-827AD Li Zhengfeng wrote in an ode to peonies, "their national colours in the morning show the heart's content during carousal, while their heavenly fragrance at night dyes dresses." This is the origin of the well-known expression "National Beauty and Heavenly Fragrance". In the meantime, there was competition to plant peonies in temples and monasteries, such as Ci'en Temple and Xingtang Temple. The people in Chang'an had an ardent love of peonies. This passion was described in Liu Yuxi's poem: "Only the peony is a real national beauty. Its blossom creates a sensation throughout the capital." Also, in *Addendum of National History*, Li Zhao of the Tang says: "One was regarded as shameful if one did not visit the flowers." Since people treated peonies with such devotion, the price rose rapidly. "People grow them for profit. One could be sold at several ten thousand coins." (*Addendum of National History* in the Tang Dynasty).

Along with peony development in Chang'-an, the plants were gradually spread to other places. During the period from 724 - 749AD, they were introduced to Japan, while within China, the plants flourished in Luoyang. In the Tang (923-926AD) the Emperor Zhuangzong constructed Linfang Hall in Luoyang and "cultivated over a thousand peonies at the front" (*Qingyi Records* by Tao Gu of the Song Dynasty), on a similar scale to cultivation at the Tang Palace in Chang'an. Peonies were also introduced to the southern part of China. When Bai Juyi took the post of prefectural governor in Hangzhou (about 821-824AD), he once saw peonies being transplanted by monks from the capital to Kaiyuan Temple *(Comments by Yunxi* by Fan Shu, theTang).

In addition Mudan and Shaoyao, with other exotic plants, were planted in large numbers in the forbidden gardens of the Bohai State then established on the Peony Garden in the northeast. This state kept an intimate connection with the Tang Dynasty (*Notes on Hearsay in*

Songmo Area), but peonies became extinct there when the state perished.

Multi-petalled forms had been bred with the development of peony cultivation in Tang. *Miscellany in Duyang Area* noted that "multi-petalled flowers were planted in front of the Palace of the Emperor Muzong" (821-824AD). Peony cultivation techniques were greatly developed in the Tang Dynasty. For the newly transplanted peony, "a cloth tent was spread out and a twig fence was woven to protect it. Because it was sprinkled with water and its root was covered with original soil before trans-planting, the colour of its flower remained as fresh as before" (Bai Juyi, *Flower Purchase*). As a result, "it bloomed one hundred percent after it had been trans-planted to hundreds of places" (Bai Juyi, *Transplanting Peonies*). According to Liu Zongyuan's *Records on the Dragon City*, "The Luoyang resident Song Danfu was skilled in the art of cultivation. He was summoned by Emperor Xuanzong, Tang, to plant ten thousand peonies in Lishan. The colours and shapes were all different." It is presumed that such variation was the result of selection after sowing the seeds of natural hybridisation.

Peony flower arranging became a grand occasion in the royal palace of Tang. The activity was very refined and the requirements were carefully prescribed: decorating environment, cuttin instruments, containers, water quality and flower stands etc. While peony arranging was being enjoyed, people would also paint pictures, play music, drink wine and write poems (Luo Qiu, Tang, *Hua Jiuxi*).

In the North Song the centre of peony cultivation shifted to Luoyang, and as a result the Luoyang peony became foremost in the country. Planting peonies and appreciating the flowers became the prevailing fashion. When the peonies were in bloom in the Emperor's garden, "tents were arranged and shops were established. Musical pipes and strings could be heard inside. Town ladies stopped their cooking to visit the flowers" (Li Gefei, the Song Dynasty, *Records on Luoyang Famous Gardens*, hereafter *Li's Records*). Thus great flower fairs and gatherings were established. It became a custom for Luoyang women, no matter how rich or poor, to wear peonies in their hair (Ouyang Xiu, the Tang, *Records of Luoyang Peony,* hereafter *Ouyang's Records).*

Grafting was then introduced to propagate new varieties selected from the continued production of seedlings. Based on this, a succession of specialised books on peonies was published. In 1034AD *Ouyang's Records* was first printed, contributing greatly to China floriculture and to the study of flowers and cultivated varieties. Ten years later *Pictures and Poetry of the Luoyang Peony*, written by Ouyang Xiu, became another key book in peony literature. After that, *Records of Luoyang Peony* by Zhou Shihou, 1082AD, supplemented *Ouyang's Records.* It should be emphasised that two strains of crown form had evolved in 50 years into crown proliferate, the same time as single into crown. The evidence suggests that gradual development and favourable conditions are the dynamics to accelerate the natural evolution process.

Chenzhou (present Huaining of Henan Province) was another area where a great number of peonies were grown in the Northern Song Dynasty. The flowers were all introduced from Luoyang. "Farmers planted the flowers in vasts area counted by *qing* (1 *qing* = 6.667 hectares), on a similar scale to that of food crop planting" (*Records on Chenzhou Peonies* by Zhang Bangji of the Song Dynasty). But this delightful scene was short-lived. When the Nuzhen nationality army invaded, all the plants were completely destroyed.

Peonies in Chengdu, Sichuan Province, began to thrive during the period between Tang and Song Dynasties. Xu Yanqiong, the maternal uncle of the descendant Emperor (on the throne 919-932AD) of the Earlier Shu State, brought a very old peony the 3000 *li* (1500 kilometres) from a Buddhist monastery in Qinzhou (present area around Tianshui, Gansu Province) to transplant it in his new residence in Chengdu. Afterwards Meng Chang, the descendant Emperor of the Later Shu State (on the throne 934-965AD) introduced many more peonies and "widely planted them in Xuanhuayan Garden, which was named Peony Garden. But the flowers then spread out to local families when the Shu State perished, and farmers grew peonies for profit by cultivating the seeds and dividing the root systems" (*Records on Peony*, Hu Yuanzhi of the Song Dynasty).

Peonies in Tianpeng (present Pengzhou, Sichuan Province) enjoyed a high reputation in the Southern Song Dynasty. Tianpeng peonies came mainly from three sources: those that accumulated in Tianpeng after the plants came into the possession of ordinary people with the Later Shu's overthrow; new varieties which were steadily purchased from Luoyang by local people; and the original native plants, most of which had single flowers. *Annals of County Peng* written in the Qing Dynasty indicated that Mount Danjing, 32 *li* north-west of the town, was named "red scenery" from the peonies abounding there. "Flowers were planted in plots on the mountain." It was also said that a peony planted there in the period of Earlier Shu was "as large as a full-grown tree".

Peonies in the regions south of the Yangtse River also developed in the Song Dynasty. Only the Preface of *Description of Peony in Yue* (Zhong Shu,986AD, *Collection of Chinese Agriculture Works*) has survived. This mentions that "In the region of Yue only peony was admired. There were 32 extremely beautiful strains" (Monk Zhong Shu of the Song Dynasty, *Preface of Description of Peony in Yue*). In *Description of Flowers in Wu* (Li Ying, 1045AD, *Collection of Chinese Agriculture Works*), 42 varieties different from those in Luoyang were noted. Peony cultivation declined in Luoyang after the Song imperial court moved to the south, but it was developed in Hangzhou, and some novel varieties were produced.

Peony production was at a low ebb during the Yuan Dynasty, leading Yao Sui to comment, in his *Peony Criticism*, "It is as hard to find a multi-petalled flower as to find a thousand heroes or ten thousand elites, only because the world cannot be kept persistently peaceful."

The peony cultivation centres moved to Bozhou (Anhui Province) and Heze (Shandong Province) in the Ming Dynasty, and to its capital Beijing, and the peony gradually began to thrive again. Also in the regions around Taihu Lake to the south of Yangtzi River, Lanzhou and Linxia in the north-west, peony planting techniques were developed and improved until the Qing Dynasty and even the Republic of China. Xue Fengxiang of the Ming Dynasty recorded in his book *Peony History in Bozhou* (1613-1617) that big families "searched everywhere in other regions for better varieties and transplanted them to Bozhou". They spared no expense for famous varieties and "frequently purchased a young seedling with one thousand coins and looked after it carefully." As a result the peony gradually prospered in Bozhou, reaching a peak of interest in the periods of Longqing and Wanli (1567-1620). The book notes 274 cultivars and describes about 200 of them in detail. Additionally, Niu Xiu noted 141 varieties in *Statements on Bozhou Peony* in 1684, the 22nd year of Kangxi, Qing Dynasty.

Caozhou was different from Bozhou in terms of new strains, and had a better reputation, so it became the peony

cultivation centre in China during the Ming Dynasty. Varieties were exchanged between the two places. "Flowers in Caozhou were always transplanted from Bozhou" (Yu Pengnian of Qing Dynasty, *Florilegium of Caozhou Peony*). On the other hand, many famous strains in Caozhou were introduced to Bozhou. Jiang Tingxi of the Qing Dynasty commented in his *Section of Prefecture of Yanzhou* that peony cultivation had been investigated. "It initially flourished in Luoxia (Luoyang), and then in Bozhou, where plants are grown singly. Caozhou was the place for grand displays of peonies." Xie Zaihang, once the chief of Dongping Prefecture, recalled the sight of peonies when in Caozhou: "For 100 *li* as I went along the road in Puzhouand Caonan, my nose was filled with fragrance, because in each family's fields, peonies were planted as widely as vegetables were cropped" (*Criticism in Five Groups*). He then appreciated the peonies of a scholar's family: "Flowers completely filled a garden of about 50 *mu*. Even a tiny piece of land beyond a pavilion was full. The sight was similar to a huge sheet of brocade, dazzlingly colourful" (ibid). "Therefore in the beginning years of the Qing Dynasty, the peony in Caonan was eulogised and enjoyed the highest fame in China" (Yu Pengnian, *Florilegium of Caozhou Peony*). Peonies flourished in Caozhou for a considerably long period, lasting through the two dynasties of Ming and Qing. Zhao Shixue in his compilation *New Additions to the Peony Florilegium* (1911) recorded 240 strains of Caozhou peonies.

Peony cultivation prospered in Beijing from the time when the city became the capital of the Dynasties of Liao and Jin. A book from the Ming Dynasty, „*Study on Beijing* indicated that an emperor of the Liao Dynasty went to Changchun Palace to enjoy the peonies. Beijing remained the capital in the Dynasties of Ming and Qing and there was a boom in peony cultivation in Ming. "Inside and outside the Royal Golden Palace peonies were planted everywhere." There are still three great gardens completely given over to peonies in the suburbs of Beijing. It is said, in *Huangdun Collection*, an anthology, that in the first, Liangjiayuan Garden, the scene might be similar to that in ancient Luoyang. The second is Qinghuayuan Garden where in the Ming Dynasty an emperor's maternal relative's villa was situated. *Records of Tourism in Capital Yan* (ie Beijing) noted "There were many rare varieties in the garden, 'Lu Hu Die' being the superior one. At flowering time the garden deserved to be called a sea of flowers." The last garden, Zhangyuan, located in present area Huayancun, belonged to a nobleman. In *Outline of Imperial Capital Scenery,* Liu Tong and Yu Yizheng noted, "When the capital's peonies came into bloom everybody would go directly to Garden Zhangyuan." Yuan Hongdao, great writer of the Ming Dynasty, once went to appreciate the flowers. In his description " approxi-mately 5 000 peonies were blooming. One particular cultivar was superior: it was as white as snow in the early morning; between 9 and 11 o'clock, tender yellow; while at noon it blushed like rosy cheeks. So it was really unique among flowers" (Yuan Hongdao of Ming Dynasty, *Records of Peonies Enjoyed in ZhangYuan Garden)*. *Annals of Shuntian Prefecture* and *Outline of Imperial Capital Scenery* further noted that in 18 villages in the areas of Fengtai and Caoqiao, peony cultivation was the main employment. "Caoqiao beyond You'anmen had spring water readily available to the north. The residents there, flower planters by profession, warmed their peonies with mild fire, so that in the middle of October flowers could be sent into imperial palaces." In the Ming Dynasty peonies were also luxuriantly planted in Jile Temple. In the beginning of the Period of Tianqi (1621-1627), "outside the gate there were old willows and in front of the temple ancient pines. On the left of it was situated the hall of National Flower Peony." This shows that the peony was honoured as the national flower early in the Ming Dynasty.

In the Qing Dynasty fresh flowers were provided to the Imperial Palace in all four seasons. Peonies became available for winter decoration when the technique of forcing was developed. By the end of the dynasty peonies had been widely planted in the Imperial Palace Garden, the National Flower Terrace of the Summer Palace and Chongxiao

Temple of Baizhifang. The National Flower Terrace, also named Peony Terrace, was initially built in 1903 to the east of the Cloud Dispelling Palace. Cixi, the mother of the emperor, issued an imperial edict declaring the peony as the national flower and ordered a stone carved with the words National Flower Terrace. Other peony growers usually purchased peonies from Caozhou, and planted some of their plants in pots, using the technique of forcing to produce flowers for the Spring Festival. The rest were grown in nursery beds to produce seedlings for sale.

Jiangyin described the thriving peonies south of the Yangtse River in the Ming Dynasty. "In the present period peonies are particularly famous in Bozhou in the north and in Jiyang in the south" (Jiao Hong of the Ming Dynasty, *Preface to Peony History*). In addition, a great number of peonies were planted in Hangzhou and Suzhou. Ji Nan in his *Peony Florilegium* (1809) summarised the experience of peony cultivation and breeding in these areas. He collected 103 varieties, 24 of which came from Bozhou, 19 from Caozhou and the rest from the area of Taihu Lake. Since grafting with shaoyao as rootstock began, peonies there "became better."

Peonies in Shanghai also once enjoyed quite a reputation. After five harbour cities were built up in the Qing Dynasty, the cultivation of peonies in Shanghai prospered. "The best ones were produced at Fahua Temple" (*Annals of Count Shanghai* written in the Qing Dynasty). This reputation however was taken over by Huangyuan Garden, where Japanese and French varieties were also introduced. All the famous traditional varieties of ancient China recorded in *Mirror of Flowers* could be found in the garden before it was unfortunately destroyed in the time of the war with Japan.

Ning'guo peony and Tongling peony were in the group of peonies south of the Yangtse River. As to the peonies grown in TongLing, the *Annals of County TongLing* stated, "the variety known as Fairy Peony grows in the mountains in rock crevices. A white-flowered peony plant, over one *chi* high with two or three matchlessly pure and pretty flowers, is said to have been grown by Ge Hong." If it is true, the plant must be 1600 years old.

In the dynasties of Ming and Qing, peonies in Gansu Province quietly prospered as well. Peonies were planted in most parts of the province. It was said that the peonies in the Taoist Temple Jintain remained from the Tang Dynasty. Tan Sitong reported: "though peonies are produced all over Gansu, those in the governmental garden were superb: all the several hundreds of plants in the garden grow to over roof height and the flowers borne on each plant are numbered in hundreds." In *Annals of Hezhou* compiled in the Ming Dynasty (1563) peony cultivation was already recorded. Wuzhen (1721-1797), a poet in Gansu in the Qing Dynasty, wrote in his poem "Peonies can be seen in every place, those in Hezhou are most praised." Another *Annals of Hezhou* finished before 1949 noted, "There were scores of varieties of peonies in the past. Because of efficient cultivation techniques, the plants in the province at the present time are superior."

Wanhua Hill in Yan'an, facing Huayuantou Village, was luxuriantly planted with peonies. In *Annals of Yan'an Prefecture* written in the period of Jiajing of the Ming Dynasty (1522-1567), peonies were recorded to be "greatly planted in Huayuantou and woodmen cut them as firewood." For a thousand years, the local people nearby have kept the tradition of strolling around the fairground in Huayuantou on April 8th of the Lunar Calendar to appreciate the flowers. In 1939 and 1940, the leaders of the Communist Party of China Mao Zedong, Zhou Enlai, Zhu De, Dong Biwu, Lin Boqu and Ren Bishi twice went to enjoy the peonies.

Peonies were produced in Guanyang of Guangxi as well. In *Complete Annals of Guangxi* written in the Ming Dynasty, it was claimed that "Peonies came from the Guangxi Counties of Lingchuan and Guanyang. Peonies in the latter place grew as high as one *zhang*, so that county was called little Luoyang."

The following conclusions can be drawn after a panoramic view of Chinese peony cultivation is taken:

First, whether peony cultivation prospers or declines is decided by the rise or fall of the country's fortunes. Of course, one could take the view that the vicissitudes of a nation are a reflection of the wax and wane of peony planting. In *Li's Records* it says, "The ups and downs of Luoyang can be read from the state of the peony gardens." The changes in gardens and flower cultivation can be regarded as a barometer of the politics and economy of a nation.

Secondly, the pattern of Chinese peony cultivation that emerges through its long history is that the main centres are along the middle and lower reaches of the Yellow River, while other areas become influential secondary cultivation centres. As dynasties change, so do the peony cultivation centres. In outline, the formation and development of Chinese peony varieties have closely followed this geographical pattern. The centres are summarised in Table 1.

Thirdly, 1 650 years have passed since Chinese peonies were introduced from the wild state into ornamental cultivation. The cultivated varieties gradually evolved from simple flowers through more complex multi-petalled forms, culminating with the crown proliferate varieties. The techniques of grafting have enabled the cultivation in great numbers of new cultivars with good properties.

Finally, frequent upsurges in the popularity of the peony reflect the esteem in which it has been held nationwide in each dynasty. It has long had national flower status. Although since Tang there have been many changes of dynasty, the celestial beauty and fragrance of the peony have sustained its exalted position. The tree peony has enjoy-ed the highest status, fame and reverence of all of China's many lovely flowers.

Table 1 Chinese Peony Cultivation Centres Through the Ages

Dynasty	Sui	Tang	The Five Dynasties	Northern Song	Southern Song	Ming	Qing
Period (AD)	581~618	618~907	907~960	960~1127	1127~1279	1368~1644	1644~1911
Main Centres	Luoyang	Chang'an	Luoyang	Luoyang	Tianpeng	Bozhou Caozhou	Caozhou (Heze)
Important secondary centres		Luoyang, Hangzhou, Mudan jiang	Chengdu, Hangzhou	Chenzhou, Hangzhou, CountyWu, Chengdu	Hangzhou	Jiangyin, Beijing, Chengdu, Luoyang, Guanyang	Beijing, jiaxing, Shanghai, Linxia, Ningguo, Lanzhou, Chengdu, Luoyang

Varietal Classification System

Flower colours, florescence and multi-petal properties were the criteria for peony classification in the past, but used in explicit and overtly practical ways. At present flower form is widely used as the basic taxonomic character, and one which is capable of demonstrating the characteristics and indicating the evolution of varieties. So it is one of the most important aspects of systematic methods of conducting production and research. The system of classification presented here is based on "The Classification of Herbaceous and Tree Peony Flower Forms" suggested in the early 1960s by the late peony expert Zhou Jiaqi. It takes advantage of extracts from other classification schemes, and was gradually constructed taking account of research in cytology, palynology and biochemistry of peony cultivars, together with practical experience.

Principles of Varietal Classification

1. The classification must be based on correct analysis of varietal characters and evolutionary indications, and have the aim of convenience and practicability. The origin, position, practical value, production and development levels of cultivars should be recorded.
2. The varietal classification should be based on the premise of the classification of species origin. This point was neglected before the 1980s, but in the last ten years the investigation, utilisation and research of peony species and their sources has been an area of increasing focus. The question of peony cultivar origin is complicated. Some are descendants from a single origin, while others have different hybridised ancestors. In addition, many varied groups of population strains appear to result from geographical location, environmental consequences and human activities, so that in the classification of peony varieties the original species should first be considered. In 1990 Qin Kuijie et al suggested the new system in which species origin is considered to be the first level of classifying criteria. That was an important development. Subsequent additions and revisions have been made according to further morphological, palynological and biochemical studies.
3. Flower form is maintained as the essential criterion because the structure and composition most readily and concentratedly demonstrate and can represent typical characteristics and differences between peony cultivars. The properties of other component parts of the plants can be regarded as secondary criteria for classification.
4. The same classifying system can be used for both tree peonies (Mudan) and herbaceous peonies (Shaoyao) in cultivar taxonomy, because they share similar flower forms. However, the most prominent flower forms among Mudan cultivars are different from those among Shaoyao cultivars.

Basis of Varietal Classification

The varietal classification system has four levels: Series (including Sub-series), Cultivar Groups, Sections (including Sub-sections) and Forms.
1. There are two series: herbaceous peonies (Shaoyao) and tree or shrub peonies (Mudan). There is effectively a third series, Itoh or interseries peonies which consists of peonies that result from crosses between Mudan and Shaoyao. These plants have been developed mostly in Japan and United States of America (where they are called inter-sectional peonies), and more recently in China.

In each series, the descendent cultivars are either from a single origin, or from crossing several cultivars, and this establishes the two sub-series under each series.
2. Different species sources are the basic criterion for classifying cultivar groups. They function essentially in the characteristics of descendent varietal populations formed after evolution and development. On the other hand, the descendent populations show hereditary characteristics of their parental origin.
3. The quantity of petals that makes up a flower form is the main basis of classifying sections. Most of the peony cultivars, about 80%, are multi-petalled. Dissection and observation reveal that there are two forms of multi-petalled flowers. One has the normal structure of individual flowers but with increased numbers of petals, the other has the appearance of single flowers combined vertically like a tower, one overlapping the other. It is appropriate to divide sections according to the degree of petal increase of flowers, and to divide sub-sections by petal origins.
4. The flower structure and degree of heteromorphosis are the main elements of the final division, that of flower form. Different flower forms of peony cultivars are determined by the amount of different petal development, evolutionary origins, and the degree of heteromorphosis. Each flower form represents a certain stage of evolution, determines the position of a cultivar and shows the characteristics of the variety. Consequently, the degree of evolution and also genetic relationships between cultivars are indicated.

Evolutionary Developments and Construction of Flowers in Different Forms

There are four evolutionary developments which are followed regularly and sequentially in peonies to construct the flower forms.

1. Natural Addition of Petals

The ancestral peony species have flowers of single form consisting of a whorl or two whorls of petals. But in artificial cultivation conditions, with crossing and selection, semi-double and full double flowers gradually appeared. The petals increase from exterior to interior and from large to small petals. The Hundred-Petal sub-section is formed from the Single-Flower section. In the course of evolution, layers of petals were increasingly added as the flower forms evolved from a lower to a higher degree: Single Form → Lotus Form → Chrysanthemum Form → Rose Form. The characteristics of this sub-section include natural and whorled increase of petals, a corresponding decrease or slight petaloidy of stamens, generally normal pistils and flat and regular flowers.

2. Stamen Petaloidy

The stamens in single flowers grow normally and completely, while those in semi-double and full double flowers show two kinds of heteromorphosis. One is stamen petaloidy, ie stamens become petaloid with the shape and colours of normal petals. The other is stamen reduction, where stamens become needle-shaped and non-functional, or totally disappear. Observations of flower bud differentiation and development indicate that the differential growth or petaloidy of the stamens is basically centrifugal. The nearer to the centre the petals of stamen petaloidy are located, the taller and bigger they are, and vice versa. Thus the Crown sub-section is formed. From different degrees of stamen petaloidy, an evolutionary path is produced: Single Form → Golden-Stamen Form → Anemone Form → Golden Circle Form → Crown Form → Globular Form. The characteristics of flower forms in this sub-section are the stability or only slight increase of original petals, and the mainly centrifugal petaloidy of stamens. Recent observations show that whorled petaloidy of stamens can be seen in some cultivars. Further research on this is needed.

3. Pistil Petaloidy

Pistil petaloidy evolution and pistil disappearance is frequ-ently and particularly found in the flower forms of the Proliferate-Flower section. Their state and degree of petaloidy varies greatly between varieties.

Some flowers show slight changes with petaloid stigmas but normal ovaries. In others both ovaries and stigmas have the shape of petals, some still preserve the yellowish-green that was the original colour of the ovary, others have striped petal colours at the apex. Other possibilities are the pistils completely petaloid with shapes and colours of normal petals, or most pistils petaloid with petal colours, keeping a trace of the ovary only in the central part. The lower degree petaloidy usually occurs in the section of single flowers, the higher degree of petaloidy in proliferate flowers. The degeneration or disappearance of the pistils can be seen in high-ranking forms of both sections.

4. Top-Bottom Overlapping in Single Flowers

One important type of the higher degree petal development has the appearance of top-bottom overlapping of two or more lower-degree flowers. It is the essential basis of the structure of Proliferate Flowers. Generally, the (apparent) overlapping top flower has a lower degree of evolution and the bottom flower a higher degree. The lower flower indicates the degree of evolution and the characteristics of the Proliferate Flower section. The forms of the bottom flower provide the criteria for classifying the Hundred Proliferate Flower sub-section and the Crown Proliferate Flower sub-section. The evolutionary sequence of the former is Lotus Form → Chrysanthemum Form → Rose Form, and of the latter is Anemone Form → Golden Circle Form → Crown Form → Globular Form.

The classification and evolution sequence of peony cultivar forms is described diagrammatically in Table 2.

The New Revised System of Cultivar Form Classification

The system consists of three series, with four sub-series; 12 or more groups; two sections, with four sub-sections; and 16 forms.

The Criteria of Level One: The Divisions of Series and Sub-series

1. Tree Peony or Mudan Series

The varietal population of the main ancestral species and their later hybrid generations are included in this series. Most of the present Chinese and foreign Mudan varieties are in this series. There are two subordinate sub-series.

(1) Single Species Tree Peony Sub-series These are cultivars which originated and have been developed from a single ancestral species. Examples are the 'Danfeng' varieties, widely cultivated and planted now for medicinal purposes, plants from the ancestral *P.ostii*, and plants from *P.rockii* such as the group of cultivars known in the western world as Rock's variety.

(2) Hybrid Tree Peony Sub-series Cross-bred from many ancestral species, for example the many varieties of theCentral Plains group, most widely planted and distributed in China. Other examples are the dark-purple varietal population Saunders series from the United States, and the mainly yellow Lemoine series from France.

2. Herbaceous Peony or Shaoyao Series

(3) Single Species Herbaceous Peony Sub-series These are cultivars which originated and have been developed from a single ancestral species, for example Chinese herbaceous peony cultivars developed from *P. lactiflora*. There are also cultivars known to originate from the group of Wittmanniana species. There is another small group of widely grown full-double cultivars whose origin is unclear but which are associated with the name officinalis, although their morphology is distinct from the widely distributed *P. officinalis* group of wild species.

(4) Hybrid Herbaceous Peony Sub-series There are many popularly-planted cultivars mostly of relatively recent origin in Europe and in America

Table 2 Peony Cultivar and Form Classification and their Evolution Sequence

```
Ancestral Species
├── HP Series
│   ├── HP Subseries
│   │   ├── CGHP
│   │   ├── CG Medicinal HP
│   │   └── CG Potaninii HP
│   └── Hybrid Subseries of HP
│       └── CG New Americ HP
├── TP Series
│   ├── TP Subseries
│   │   ├── CG Rockii TP
│   │   └── CG Ostii TP
│   └── Hybrid Subseries of TP
│       ├── CG Northwest TP
│       ├── CG Central Plains TP
│       ├── CG South Yangtse TP
│       ├── CG Southwest TP
│       ├── CG French Lemoine TP
│       ├── CG American Saunders TP
│       └── CG Japanese TP
└── Hybrid Series of HP and TP

Single Flower Section
├── Hundred Petals Subsection → Single F → Lotus F → Chrysanthemum F → Rose F
└── Crown Subsection
    ├── Golden Circle F
    └── Golden-Stamen F → Anemone F → Crown F → Globular F

Proliferate Flower Section
├── Hundred Proliferate Flower Subsection → Lotus Proliferate F → Chrysanthemum Proliferate F → Rose Proliferate F
└── Crown Proliferate Flower Subsection
    ├── Anemone Proliferate F
    └── Golden Circle Proliferate F → Crown Proliferate F → Globular Proliferate F
```

Key: HP = Herb Peony, TP = Tree Peony, CG = Cultivar Group, F = Form

that have been developed by crossing ancestral species and their cultivars which are classified in this group.

3. The Herbaceous Peony and Tree Peony Hybrids Series

This series includes the descendent varietal population of the hybrids between species, or between species and cultivars, of tree peonies and herbaceous peonies. Examples are a hybrid yellow peony produced in Japan by crossing the herbaceous peony Huaxiangdian and a tree peony Jinhuang of the Lemoine cultivar group. Also, Itoh type hybrids (called intersectional hybrids in the United States of America). The introduction of these plants was a sensational development in the world of peonies. Plants in this series have been produced in China recently. At present there is not a stable varietal population of a significant size and they have not been widely used. Neverthe-less it seems appropriate to assign a position for such plants in the classification, and series is perhaps the most appropriate level.

The Criteria of Level Two: The Divisions of Cultivar Groups

1. The single species Tree Peony subseries showing clear species origin and applied in production

(1) Cultivar Group of Rockii Tree Peonies All show the ancestral features of *P. rockii* : a large striking blotch of dark purple or purple-black at the bottom of the petals, milky white stigmata, filaments and flower discs, narrow leaves and leaflets.

(2) Cultivar Group of Ostii Tree Peonies All show the definitive features of *P. ostii* , the varietal population of 'Fengdan' cultivars grown mainly for medicinal purposes. They have tall and straight stems, long internodes, narrow leaves with few incisions, flat and regular flowers.

2. Hybrid Tree Peony Series

(3) Northwest China Tree Peony Cultivar Group A population of hybrid descendants mainly from *P. rockii* and also *P. spontanea*. They have an obvious purple blotch at the base of the petals. Some show the features of *P. spontanea*, others the characteristics of *P. rockii*. In Gansu, north west China, most cultivated varieties are categorised in this group, which has over 100 cultivars.

(4) Central China Plains Tree Peony Cultivar Group A hybrid population developed mainly from *P. spontanea* with *P. Rockii* and *P. ostii* influences. The varietal population of earliest horticultural interest.

(5) Southern Yangtse Tree Peony Cultivar Group A hybrid descendent population formed from the crossing of *P. ostii* and the Central Plains peonies that were moved south in ancient times before they adapted to the local environment.

(6) Southwest China Tree Peony Cultivar Group This group shows a complicated mixture of resource origins. The small number of varieties and scattered distribution of cultivation necessitates further research and the reservation of its category position.

(7) French Lemoine Tree Peony Cultivar Group A hybrid descendent population of mostly yellow-flowered plants derived mainly from crossing *P. delavayi* and *P. lutea* and the ancestral varieties of *P. spontanea*.

(8) American Saunders Tree Peony Cultivar Group A hybrid descendent population including yellow- and black-purple-flowered plants from crosses between *P. delavayi* and *P. lutea* and Japanese tree peony cultivars.

(9) Japanese Tree Peony Cultivar Group Although the tree peonies in Japan were introduced from China in Tang, the breeding of improved cultivars during the period 1603 - 1867 produced a specialised and Japanese style of Mudan. The typical Japanese tree peony varietal population that gradually evolved is mainly characterised by slightly multi-petalled flowers with thick petals and bright colours, flat and neat flower forms and stiff flower stalks.

3. Herbaceous Peony Sub-series
The following sub-series can be identified:
(10) Chinese Herbaceous Peony Cultivar Group The varietal group population developed mainly from *P. albiflora* in China, Japan and Korea, historically the earliest horticultural varieties.
(11) European Herbaceous Peony Cultivar Group The varietal group population developed from the complex species groups of *P. officinalis* and *P. mascula* used originally as medicinal plants.
(12) Eurasian Herbaceous Peony Cultivar Group The varietal group population developed from species with finely divided leaves such as *P. anomala* and *P. tenuifolia*.

Clearly, other cultivar groups which would belong in this sub-series could be identified, for example cultivars derived from the Wittmanniana group of species.

4. Hybrid Herbaceous Peony Sub-series
During the twentieth century a great many cross-breeding experiments within the herbaceous peony series were carried out by European and American thremmatologists, using a very wide variety of species and hybrid cultivars. The descendent populations have greatly enriched the range of cultivars. Aggregate varietal groups of European and of American cultivars can be defined but the only identifying feature is great diversity, so apart from historical interest the usefulness of such groups is very limited.

The Criteria of Level Three: The Divisions of Sections and Sub-sections

Two sections, of single flower and proliferate-flower, and four sub-sections, are defined according to the extent and structure of petal proliferation and the sources of petals.

1. Single Flower Section
Flowers whose appearance or structure is that of an individual flower
(1) Hundred Petals Sub-section
Naturally increased normal petals gradually smaller towards the centre of the flower. With the increase in the number of petals, there is a corresponding decrease or even disappearance of stamens, or possibly stamens only slightly petaloid in the centre of the flower. The pistils are normal or sometimes reduced. Flowers are flat and neat and generally regular.
(2) Crown Sub-section Here stamen petaloidy is the main source of petals. There are 2 - 4 whorls of wide and straight exterior petals, and slender wrinkled or curled interior petals which are mainly developed from stamens as they become petaloid from the centre outwards. Pistils are normal, or petaloid, or reduced, or completely disappeared.

2. Proliferate Flower Section
Flowers whose appearance or structure is that of more than one flower combined together
(3) Hundred Proliferate Flower Sub-section Here the flower form has the appearance of top-bottom overlapping of two or more individual flowers of the Hundred Petals sub-section.
(4) Crown Proliferate Flower Sub-section Here there are various forms which have the appearance of two or more individual flowers of the Crown sub-section overlapped from top to bottom.

The Criteria of Level Four: The Divisions of Forms

Different forms within the divisions of Level Three are defined by different amounts of petals or different degrees of petaloidy.

1. Flower Forms in the Hundred Petals Sub-section
(1) Single Form 1 - 3 whorls of normal petals that are wide, large and flat with a wide-ovate, ovoid or obovate shape. Stamens normal and pistils normal and fertile.
(2) Lotus Form Large and neat petals in 4 - 5 slightly overlapping whorls forming the shape of a lotus flower. Stamens normal and pistils normal.
(3) Chrysanthemum Form 6 whorls. Petals gradually decreasing in size towards the centre. Stamens normal or fewer and petaloid in the centre of the flower; pistils normal.
(4) Rose Form Petals generally longer than in Chrysanthemum form, becoming smaller from the outside towards the centre. Most stamens disappeared. Pistils normal, slightly petaloid, reduced or completely disappeared.

2. Flower Forms in Crown Sub-section
(5) Golden Stamen Form 2 - 3 whorled petals, large and straight. Bright golden stamens, larger anthers and filaments. Pistils normal.
(6) Anemone Form 2 - 3 outer whorls of wide and straight petals. Stamens completely petaloid and have become narrow and straight petals. Pistils normal or reduced.
(7) Golden Circle Form 2 - 3 outer whorls of wide and large petals. Most stamens petaloid, but a whorl of normal stamens remains as a golden circle between the narrow interior petals and

the wide outer petals. Pistils normal, or petaloid, or reduced.

(8) Crown Form Wide and expanded outer petals. Completely petaloid stamens usually with the appearance of becoming larger from outside to inside, sometimes mixed with a few narrow silk-like incompletely petaloid stamens. Pistils are petaloid, reduced or completely disappeared. The centre of the flower is raised, forming a crown shape.

(9) Globular Form All stamens are highly petaloid with shapes and sizes similar to those of normal petals. Pistils petaloid or reduced. The whole flower resembles a Chinese artistic ball.

3. Flower Forms in the Hundred Proliferate Flower Sub-section

(10) Lotus Proliferate-Flower Form The flower structure is that of two individual flowers of lotus form overlapping. Only a few examples in tree and herbaceous peonies are known.

(11) Chrysanthemum Proliferate Flower Form The flower structure is that of two individual flowers of chrysanthemum form overlapping. The pistils in the bottom flower are slightly reduced and growing around the top flower. The top flower has fewer petals, but stamens and pistils are generally normal.

(12) Rose Proliferate Flower Form The flower structure is that of two individual flowers of rose form overlapping. In the bottom flower the petaloid pistils have many shapes and colours. Some even have the appearance of normal petals. Different varieties show different degrees of petaloidy. The top flower does not develop as well as in the Chrysanthemum Proliferate Flower form. There are fewer stamens or slight petaloidy; pistils are slightly petaloid or degenerate and smaller.

4. Flower Forms in the Crown Proliferate Flower Sub-section

(13) Anemone Proliferate Flower Form The form is that of top-bottom overlapping of two individual anemone form flowers. This flower form represents a small proportion of both tree and herbaceous peonies. The top flower does not grow much, so that a few stamens are normal or degenerate and smaller pistils often remain in the flower centre.

(14) Golden Circle Prolifeate Flower Form The form is that of top-bottom overlapping of two individual flowers of golden circle form.

(15) Crown Proliferate Flower Form The form is that of top-bottom overlapping of two individual flowers of crown form with strong growth and various colourful petals. A few stamens remain in the centre of the top flower; pistils are degenerate and reduced or disappeared.

(16) Globular Proliferate Flower Form The form is that of top-bottom overlapping of two individual flowers of globular form, both growing strongly. In the lower flower the petals from petaloid stamens and pistils have widened to the same size as normal petals. Those in the top flower are petaloid or degenerate. The whole flower has the shape of a ball.

Flower Forms From Section Level in the New Revised System of Varietal Classification

Figure 11 illustrates diagrammatically axial sections of flower forms.

Several Points about the New Revised System

1. The system is available for the classification of flower forms of all horticultural cultivars in the series of herbaceous peony, though it has been developed from research on Chinese tree peony varietal populations. Despite the different biological characteristics, climatic conditions and cultivation histories of Mudan and Shaoyao, they share similar chromosome caryograms as well as similar flower construction and patterns of evolution. In addition, the ancestral characteristics described above of cultivar groups of Mudan can be more or less seen in the horticultural varietal populations of Shaoyao. The formation of the flower forms of varied cultivar groups, therefore, can be expected to be generally the same. The only differences are shown by the different degrees of evolution and the range of flower forms.

2. When the flower form of a cultivar is to be determined, the determination should be based on its most stable flower form at the highest level. Other flower forms at lower level that simultaneously appear are in the same flower form evolutionary series. For example, only Crown, Golden Circle, Anemone or Single form in the Crown sub-section can be found in plants of Crown form varieties. The flower forms of Hundred Petals sub-section will not appear there and Globular form, a degree higher than Crown form, also will not appear. If this is not so, the determination of the flower form is incorrect, or the cultivar has probably developed to a higher level. Consequently continuous careful observation must be carried out.

3. The classification system is, after all, an artificial classification though it objectively involves biological characteristics of cultivars and to some extent the evolutionary paths of flower forms. However biological evolution is complex and in one sense continuous, so it does not easily conform to clear divisions. Some plants with transitional flower forms between those of Hundred-Petals and Crown sub-sections do occur. For such plants it is appropriate to indicate that it is a transitional form and the sub-section that the flower form has the greatest tendency to be in.

4. Other names have been used for various of the flower forms defined above: "silk ball" corresponds to Globular; "osmanthus-holding" corresponds to Anemone; "pavilion" corresponds to Crown Proliferate; skyscraper corresponds to Hundred Proliferate.

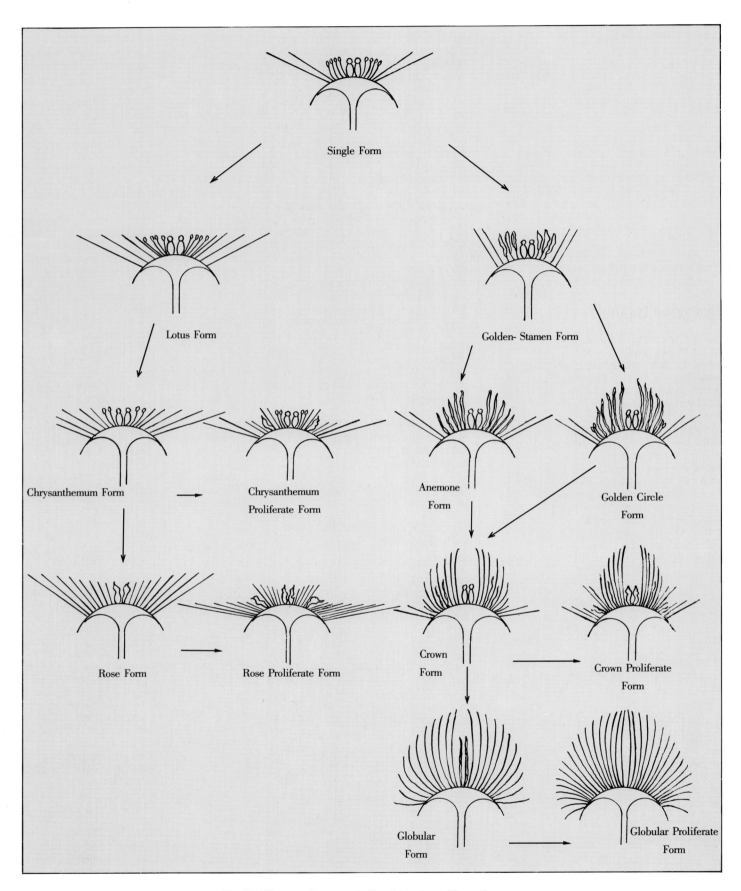

Fig. 11 Illustrates diagrammatically axial sections of flower forms

Ecological Habitats and Biological Characteristics

Ecological Habitats

All the ancestral Chinese tree peonies, ie the entire group of wild species tree peonies, are categorised as typical temperate zone woody plants. Their distribution is concentrated on the southwestern mountains and the central northern Loess Plateau and hills, forming a narrow zone of distribution. Although the southwestern area is located within warm temperate and sub-tropical zones, peonies there all grow above 2 500 metres. Elsewhere they have long been established in a temperate environment. In both areas therefore peonies have adaped to commonly occurring ecological habitats. They tend to grow better in mild and pleasantly cool climates and they have a certain tolerance of cold. They flourish in dry and high places but are less successful in hot, damp conditions. They need sunny days and yet tolerate half shade well. The cultivated varieties of Chinese peonies also show wide ecological adaptability. The distribution of the ancestral wild species is generally con-centrated in relatively restricted areas, often adjacent to and even overlapping one another. Nevertheless, the differences between species origins, elevations of individual locations, and specific environments and conditions have helped to form various cultivar groups that are able to adapt to their own ecological environment. Thus cultivar groups have developed their own particular ecological predilections. The comparison and contrast of the main meteorological elements in the principal distribution areas of each ancestral species and the cultivated areas of each cultivar group confirms this observation.

The analysis in Table 3 indicates that there exist two basically different ecological types among the wild species. The first group includes *P. spontanea, P. rockii* and *P. Ostii* with distributions below 2 000m, an annual average rainfall of less than 1000mm, maximum temperature above 30°C and minimum temperature as low as -19°C. These species share similar cool and dry ecological habits. They have a fairly high tolerance of cold and drought and can endure torrid summer heat. The other type contains *P. delavayi, P. lutea* and *P.dcomposita*. They are distributed in places where the elevation is 2000m and above, the annual rainfall is more than 1 000mm, maximum temperature less than 30°C and minimum no lower than -10°C. Their ecological predilections are similar to the previous group. However, they grow better in a warmer and moister climate and generally have a lower tolerance of cold, drought and summer heat. Table 4 illustrates a similar point, that the cultivated varities of Chinese peonies not only maintain the predilections of their ancestral parents, but adapt to a wider range of ecological conditions. In the long course of natural evolution and artifical hybridisation, cultivation and selection, cultivar groups with various ecological predilections have gradually formed, each with its own optimum ecology.

Consider first the Central Plains cultivar group. The main meteorological elements of its central cultivation areas, Heze, Luoyang and Beijing, indicate a warm temperate zone: dry and windy in spring, hot and rainy in summer, cool, fine and dry in autumn. Corresponding ecological predilections, therefore, pertain to this group of cultivars. They flourish in a temperate climate and are able to tolerate cold to some extent. They tolerate fierce heat badly. Dry, high terrain is suitable for them, but not high humidity and waterlogging. They are the warm and dry climate cultivar group.

In contrast, strong adaptability is to be expected of the Northwest cultivar group because of the meteorological elements in their main areas of cultivation. They are more tolerant than the Central Plains group of cold and dry conditions. So they can be regarded as the cool and dry climate cultivar group.

The Southern Yangtse cultivar group grows principally in Ningguo and TongLing in a zone of low mountains and hills in South Anhui. The climate is sub-tropical, fairly warm and humid in winter, hot and wet in summer with high humidity, as much as 80%. The peonies there flourish in this climate. Flowering begins early and leaf-fall is late. So this cultivar group has adapted to an ecosystem cha-racterised by high temperature and humidity.

The Southwest cultivar group contains relatively few cultivars, and their genetic relationships are unclear. They are characterised by lower tolerance of cold and strong sunshine, and need high humidity and slightly acid porous soil. These cultivation requirements result from their adaptation over time to the environment of Pengzhou and Lijiang. This is the cultivar group of warm and moist mountainous areas.

To summarise, the leading elements affecting the normal growth and development of Chinese peonies are temperature and moisture, especially humidity. The various species and cultivar groups generally react adversely to what is for them an unsuitable environment, and growth is hindered. For example, when cultivars of the Southwest cultivar group are grown in the Central Plains area, normal development cannot be relied on, and unsuccessful production of flower buds

Table 3 Main Meteorological Elements in Main Areas of Distribution of Tree Peony Ancestral Species

Ancestral Species	Area of Distribution	Altitude (m)	AvAnT (°C)	MaT (°C)	MiT (°C)	AvAnP (mm)	FS (days)	Soil and pH Value	Notes
Spontanea	Jishan	1220~1470	13~13.8	35.5	-23.2	483.3~553.9	205~222	Mt cinnamon 5.8~6.2, black mature, grayish-yellow spongy	
	Yan'an	910	9.4	39.7	-25.4	550	180		
Rockii	Shaanxi Gansu Loess plateau FA	1300~1510	7.4~8.5	36.7	-27.7	500~620	110~150	grayish-yellow spongy, cinnamon, Mt brown earth pH 5.6~6.3	
	Mt Qinling Bashan	1565~2019	6.9~15	31~43.6	-15~-22	620~834	154~220		
	Shennongjia FA	938~1200	7.4~12.2			1000~1700	153~224		
Ostii	Yangshan FA, Henan	1300~1650	12	43.6	-19	821.9	170	Mt brown earth pH 6.4	SE Gansu, SW Hubei, NW Hunan, S Anhui
Szechuanica	Maerkang-Jinchuan, Sichuan	2600~3100	8.6	34.3	-17.5	753	120	Mt yellow pH 6.4	
Delavayi	Lijiang Basin, Yunnan	2415.9~2750	9.7~12.6	26.8	-9.6	1707.6		Mt red earth	
	Yulong Snowy Mt, Yunnan	3240	5.5	18.8	-11.8	1587.5			
Lutea	Kunming, Yunnan	1980	14.7	31.5	-5.4	1006.5	341	Mt red earth, yellow earth pH 4.9~5.7	
	Huadianba, Yunnan	>2700	22	30	-3.0	1100	300		
	Linzhi, Tibet	2700~3300	9.3	15.8	0.4	650	180		

Key: FS = Frostfree Season, Mt = Mountain, FA = Forested Areas, AvAnT = Average Annual Temperature, MaT = Maximum Temperature, MiT = Minimum Temperature, AvAnP = Average Annual Precipitation

Table 4 Meteorological Elements in Main Productive Areas of Chinese Tree Peonies

Cultivar Group	Cultivation Area	Latitude	AvAnT (°C)	MaT (°C)	MiT (°C)	AvAnP (mm)	FS (days)	AvAnS (hours)
Central Plains	Luoyang	34°40'	14.5	42.5	-18	610.2	220	2200~2300
	Heze	35°20'	13.06	37.1	-19.8	706.6	212	2579.0
	Beijing	39°56'	12.2	36.78	-14.8	603	190	2780.2
Northwest	Lanzhou	36°	9.1	36.8	-21.7	327.7	130	2607.6
	Lintao	35°56'	7.0	34.6	-29.6	565.2	163	2437.9
	Linxia	35°35'	6.8	36.2	-27.8	501.7	162	2567.8
Southwest	Pengzhou	30°59'	15.6	36.9	-6.2	1225.7	277	1239.0
	Kunming	25°	14.7	31.5	-5.4	1012	341	2481
	Dali	25°43'	15.1	34.0	-3.0	1078	230	2300.6
	Lijiang	27°	12.6	30.6	-7.0	1900	285	2530
South of Yangtse River	Shanghai	31°1'	15.7	38.9	-10.0	1120	230	2020.7
	Hangzhou	30°19'	16.4	39.9	-9.6	1601.7	245	1903.9
	Tongling	30°57'	16.2	40.0	-11.9	1300~1400	237	2000
	Ningguo	30°32'	15.2	40.0	-11	1660	226	2502.7

Key: FS = Frostfree Season, AvAnS = Average Annual Sunshine

due to summer heat and dry air often leads to a lack of flowers. Exceptionally, some cultivars can adapt and bloom normally. Most cultivar groups of Chinese peonies, however, are very adaptable and tolerant of strong sunshine and different soil conditions. Most varieties benefit from moderate sunshine and at the same time are tolerant of half shade. When in bloom, lateral shading is beneficial and prolongs flowering. Shading is especially necessary for the cultivars less tolerant of sunshine. For cultivation of the Southwest cultivar group in northern areas, shading from the midday sun is appropriate, especially in summer.

All the cultivar groups have long fleshy roots, so that porous, fertile and deep soil on high and dry terrain is the most suitable. Mudan can grow well in soil that is generally not compressed and waterlogged, and a sandy soil is preferable to a sticky soil. The ideal pH is neutral or slightly alkaline.

Biological Characteristics

1. Growth Variation in Different Periods

In their long natural development and in artificial cultivation and selection, Chinese peonies have gradually formed certain biological characteristics. In the complete life span of an individual plant, there are gradual changes in their annual life cycles.

(1) Life Cycle Changes The length of the complete life span of a Chinese peony is affected by habitat conditions and cultivation in addition to genetic factors. The life span in suitable circumstances can reach several hundred years. In various places in China, there are quite a few very old peonies still living. For example, some ten peonies in the Dry Twig Peony Garden are already 700 years old. These are the oldest known in China. Among them the largest one is 2.3m high and 3m crown across. Their flowers are single form in two colours, red and white. Six ancient peonies, 500 years old, are growing in the Temple Puning in Zhejiang. They remain from the Ming Dynasty. Other peonies aged 400, 300 or 100 years are also found in China.

The different stages in the life span of Chinese tree peonies is similar to that of other trees, and in the absence of scientific evidence is divided as follows: juvenile stage, 1-3 years; adolescent stage, 4 - 14 years; post-adolescent stage, 15 - 40 years; and senescent stage, over 40 years. Generally, peonies grow slowly during the juvenile stage. One-year-old seedlings are less than 10cm high and their root system extends only about 10cm. After three years their rate of growth gradually increases. They generally do not flower until the fourth or fifth year. During the adolescent and post-adolescent stages, they grow strongly and produce many flowers. These stages are perhaps the best for appreciating Mudan. A proverb says "Old plum trees and young peonies" to indicate their prime times. In the garden of the Linxia administrative government, a postadolescent *P. rockii* has reached the height of 1.83m and is 12.6m crown across. As it produces over 300 flowers, perhaps the most in China, it is worthy of the name of Peony King".

(2) Annual Growth Cycle The annual growth cycle of Mudan, like other deciduous shrubs, shows obvious alternation between growth and dormancy. The geographical area, the year, the cultivar group and the particular cultivar, are all factors that cause variations in the annual growth cycle.

In the middle and lower reaches of the Yellow River, active growth generally begins between mid-February and mid-March. Leaf expansion occurs from the last ten days of March to the first ten days of April. The flowering period is the end of April to the middle of May. From June to October is the flower bud development phase. Leaves fall from the last ten days of October to mid-November. The period of dormancy begins when low temperatures and severe winter weather come, and lasts until the next spring.

Because of the colder weather in northwest areas, the growth period is comparatively short and the dormant period is slightly prolonged, with dormancy beginning at an earlier date.

In northeast areas such as Harbin the phenological variations are accentuated: the growth period is further reduced while the leaf fall and dormancy periods start even earlier.

In addition, the changes of annual growth patterns vary according to changes of climatic conditions, particularly temperature and air humidity, in different years. For example, in a normal year, new growth begins in peonies of the Central Plains cultivar group when the temperature is stable at 3.5 - 6°C. New twigs develop at about 6 - 8°C. At 10 - 16°C flower buds grow rapidly. The flowers bloom at 16 - 22°C, and seed is set at about 26 - 28°C. If a spell of warm spring weather occurs, the various growth phases will begin earlier. In 1993, peonies in Luoyang came into blossom seven days in advance because of the warm spring, so that the early flowering varieties had nearly faded when the peony fair was inaugurated on April 15. Careful study of these growth cyclevariations will assist in the efficient cultivation of Mudan.

2. Other Growth Characteristics

(1) Epicotyl Dormancy A dormant period of epicotyl growth occurs in all Chinese tree peonies. After seeds ripen and are sown in the same autumn, their radicle develops downwards only, into roots. Their hypocotyl stays in a dormant state. After about 60 - 90 days of low temperature, about 0 - 10°C, the epicotyl sprouts and emerges above ground level. This process can be induced by artificial methods, but the details of such methods and the phenomenon itself require further research.

(2) The Phenomenon of Dried Tips of Twigs Mudan flowers grow each year from the tips of new twigs that have developed earlier in the same year. The annual growth length of new twigs differs according to the cultivar, the age of the plant and cultivation conditions. The different lengths, from 50 to 60cm down to as low as 10cm, greatly affect overall plant total height and flower prominence. However, the length of new growth produced is not the effective annual growth, because of the phenomenon of dried tips of twigs. The top part of the year's new growth withers, becomes brown and dies during late autumn the same year. The true yearly growth is the bottom part of the new twig. This is characteristic of sympodial, as opposed to monopodial, growth. Without a terminal bud with the potential of a new shoot, the top of the annual growth does not become lignified. Only the new growth at the bottom of the new twig as far as a bud in a leaf axil becomes completely lignified. Thus the top part of the branch dies in cold winter weather and the bottom part of the branch will continue to grow in the next year. This genetic characteristic of tree peonies is possibly a result of their adaptation over a long time to cold weather of plateaus and mountainous areas. The proportion of new twig that dies varies between cultivars and with external environmental conditions.

(3) Bud Properties and Differentiation Characteristics Buds on branches are divided according to their location into terminal buds, lateral buds and adventitious buds. Generally terminal and lateral buds are formed on the annual new growth or on new shoots that appear at the base at or from below ground level.

Adventitious buds are those that develop on branches that are two or more years old. Buds can differentiate into flower buds provided they develop sufficiently. The buds that produce flowers, however, are always

Table 5 Flower Bud Differentiation Process of Some Cultivars of Central Plains Cultivar Group (Wang Lianying)

Name of Cultivar	B of FBD	A of HP	A of SP	A of PeP	A of StP	A of PiP	A of StPe	B of PiPe
SiHeLian (Lotus)	12th June	18th June	7th July	18th July	18th Aug.	11th Sept.	1~10 Nov.	
Er Qiao (Rose)	22nd June	4th July	26th July	18th Aug.	10th Oct.	2nd Oct.	1~10 Nov.	
Zhao Fen (Crown)	14th June	22nd June	12th July	26th July	16th Sept.	21st Sept.	early Jan. next y.	
Sheng Dan Lu (Crown)	6th June			12th Aug.	6th Sept.	20th Sept.	Mid Oct.	25th Nov.

Key: B = Beginning   A = appearance

Table 6 Flower Bud Differentiation Process of Some Flower Form in Northwest Cultivar Group (Li Jiajue)

Flower Form	Before D	B D	HPD	SPD	PePD	StPD	PiPD	StPe
Single Flower	20~31 May	Mid June	Mid July	1~10 Aug.	Mid Aug.	1~10 Sept.	Mid Sept.	
Chrysanthemum	20~31 May	Mid June	20~30 June	Mid July	Mid Aug. ~ Mid Sept.	Mid Sept.	1~10 Oct.	
Crown	Mid May	1~20 June	Mid June	Mid July	20~31 Aug.	Mid Sept.	20~30 Sept.	1~10 Oct.

Key: D = Differentiation

terminal and lateral buds measuring 0.5cm × 0.3cm or larger. Buds smaller than this remain in leaf bud state and do not differentiate into flower buds. The phases of flower bud development start in summer. Although the phenological periods differ between various cultivar groups, the differentiation process and the factors which regulate it and the subsequent development of buds are similar. As examples, the differentiation processes in the Central Plains cultivar group and the Northwest cultivar group are considered.

(i) Flower Bud Differentiation Process (FBD)

See Tables 5 and 6.

In Mudan varieties with relatively few petals, such as single, lotus and golden-stamen forms, the differentiation process is comparatively rapid, taking approximately 3 - 3.5 months. On the other hand, in varieties with a high degree of petal development, such as rose and crown forms, the differentiation process is much slower, taking 7 - 8 months. However, for all forms the sequence of flower bud differentiation is generally the same: flower primordium → hypsophyll pri-mordium (HP) → sepal primor-dium (SP) → petal primordium (PeP) → stamen primordium (StP) → pistil primordium (PiP) → stamen petaloidy (StPe) → pistil petaloidy (PiPe). In the flower buds of a flower of proliferate form the differentia-tion in what is apparently the bottom flower precedes the differentiation in what is apparently the upper flower.

(ii) Characteristics of Flower Bud Differentiation

It is observed that flower bud differentiation and development in both cultivar groups is basi-cally characterised as follows:

a. The differentiation and subsequent development from the flower primordium of each organ follows the localised differentiation principle and the above sequence. The degree of differentiation and development and the stability of development of each flower organ shows dramatic variation between different cultivars, particularly in petals and stamens, less so in sepals and pistils. The variations can be generally classified into two types. One is the addition of petal layers and the reduction of stamens in the differentiation process. Stamens can remain unchanged or can become petaloid in the later period of differentiation, in the cultivars 'Er Qiao' and 'Chun Hong Zheng Yan' for example. The other is the stability of petal quantity and the obvious increase of petaloid stamens in the cultivars 'Yao Huang' and 'Zhao Fen' for example.

b. The manner of petal primordium differentiation differs from that of stamen primordium differentiation. The petal primordium develops centripetally or from exterior to interior, layer by layer tangentially. The stamen primordium generally becomes petaloid and develops centrifugally, although there are reports that stamens can differentiate centripetally. Further observation and research are needed.

c. In Proliferate form flowers there is a single primordium in the flower bud from which the flower develops with the appea-rance of two flowers overlapping top to bottom. Their differentiation sequence is the same as that of Single flowers, but they differentiate from bottom to top. This phenomenon occurs more frequently in the Central Plains cultivar group than the Northwest cultivar group.

The differentiation character-istics correspond closely to the construction and evolutionary sequence of flower forms discussed previously. This suggests, therefore, that the characteristics of flower bud differentiation provide a material and theoretical basis for flower form evolution. Also it supports the flower form classification system of peony cultivars.

(iii) Characteristics of Branch Development

After the formation of flower buds, the breaking of dormancy followed by smooth and normal branch development and flowering is affected by many factors. These include not only the age of the tree and external envi-ronmental conditions including nutrient states, but also the position of buds on branches.

The phenomenon of apical dominance where the terminal bud grows faster than, or even inhibits the development of, lateral buds, exists to some extent in most plants, but is not notable in shrubs and mono-cotyledons. In areas around Beijing, apical dominance in peony growth and its absence have been observed. This may suggest that cultivars differ in their ability to generate new branches and in their flower survival rates, and that a better understanding of apical dominance may be important in planning appropriate cultivation techniques.

(4) Characteristics of Flowering
Chinese Mudan flower over many years without a break, from a seedling aged 4 - 5 years to senescent plants aged over hundred years. In a normal year it takes 50 - 60 days from breaking of bud dormancy to flowering. This period varies between different cultivars. Those flowering early or with simpler flower forms take a shorter period, but late flower-ing cultivars or those with higher rank flower forms take longer. The flowering period of different varieties is generally 3 - 10 days for individual flowers. The overall flowering period for a mature plant is about 25 - 30 days. The date flowering begins, the length of the flowering period and flower quality depend on the variety, weather conditions and the cultivation environment. A warm spring advances flowering, while a cold spring delays it. A warm winter affects some varieties that benefit from low temperature and results in poorer flowering, and is particularly harmful for plants being forced. An excessively cold winter can damage buds and thus affect flowering in the next spring. At flowering time, dry and hot winds or excessive rain shorten the flowering period and damage flowers.

Additionally, there are another two significant characteristics of Chinese tree peony flowering that should be mentioned:

(i) Concentration of cultivar flowering periods

There are over 500 cultivars in the Central Plains cultivar group of which the early and late flowering cultvars make up only 20%. The remaining 80% flower during the middle period, beginning at the middle of April and ending at the tenth day of May.

According to the statistics in Li Jiajue's *Linxia Peony*, the North-west cultivar group includes 70 cultivars of which 43, about 60%, flower during the middle period, 13 cultivars, about 20%, are early flowering, and 14, about 20%, are late flowering. Overall, 60% - 80% of cultivars flower completely in less than one month. Although this presents a splendid scene, the concentra-tion of flowering creates the impression that the flowering period is short. In future cultivation and selection, therefore, emphasis on early and late flowering cultivars is important.

(ii) Abnormal Heteromorphosis

Flower forms are the main criteria for classifying peony cultivars and, as mentioned earlier, they reveal the characteristics of cultivars and indicate their degree of evolution. Stability and precise conformity to typical flower forms, therefore, have a certain importance. The flowers of some cultivars are observed to be unstable and of variable form. Also the number of flowers fluctuates, similar to the "on and off year" phenomenon of fruit trees. For example, the cultivars 'Zhao Fen' and 'Shan Hua Lan Man' produce many flowers of large size and fully typical flower form in an on year . In contrast, there are fewer flowers and of a smaller size and irregular form in other years. Sometimes two or three flower forms of adjacent ranks are formed on the same branch, and sometimes flowers of intermediate form are produced. Stamen petaloidy is not stable in some cultivars. This is especially noticeable in the cultivars of Crown form, such as 'Yao Huang' and 'Tian Xiang Zhan Lu'. When their stamens are completely petaloid, their flowers have a full and raised centre and are typical Crown form. But sometimes stamen petaloidy is not complete; petaloid stamens may retain malformed anthers at the tip and may be narrow and short. Flowers may be small and irregular, and the colours faded. All these effects greatly reduce the quality of the flowers.

Other cultivars clearly show local adaptability in their flower forms: they present different properties in different areas even though the environmental and cultivation conditions are basically similar. For example, 'Shi Ba Hao' develops many full and splendid flowers of its typical form in Heze but not in Beijing; 'Kun Shan Ye Guang' grows well in Luoyang but less well in Heze. The phenomenon of abnormal flowering is presumably related to cultivars genetic make-up and to environmental conditions, but further observation and research are needed for it to be better understood.

Propagation and Cultivation

Propagation

Methods for the propagation of peonies include division, grafting, seed, layering and cuttings. The first three are those most frequently used.

1. Division

A simple method with a high survival rate and vigorous new plants. Although after division the new plants flower early, it takes two to three years before they completely show their cultivar characteristics. This method of clonal propagation directly maintains the advantageous features of the cultivar but the rate of propagation is relatively low.

(1) Division Time Division is carried out mainly in autumn. It is sometimes continued in spring after an early winter.

(2) Division Method A healthy plant 4 to 5 or more years old is dug up and the soil at its roots removed. After observing the structure of its branches, buds and root system, and following its natural growth lines, it may be broken into smaller plants, taking care not to damage twigs and buds. If the base at ground level is very compact, a knife or a pair of secateurs, or even a small saw, can be used to cut it, taking care to keep wounds as small as possible for more rapid healing. The number of pieces obtained depends on the size and the manner of growth of the root system. Generally, 2 or more divisions can be removed from a stock plant without digging it up, by excavating soil around the roots and breaking pieces off. After division, pruning should be carried out according to the growth of new shoots at or below ground level. If there are no such shoots, one or two adventitious buds or axillary buds on the lower part of branches should be maintained and the top parts of the branches, more than 40 - 50cm above the root, cut off. If there are two or three new shoots, the top of the plant can be cut off 3 - 5cm above ground level. Broken and diseased parts of the old roots should be removed. It is better to remove old roots if there are plenty of new roots. If there are not enough new roots, healthy older roots should be kept. To discourage pathogenic bacteria, 1% copper sulphate or other appropriate prophylactic can be used to sterilise the wounds before planting.

2. Grafting

By this method, the characteristics of the cultivar to be propagated are exactly maintained at a low cost and at a high propagation rate. It is particularly appropriate for propagating cultivars with slow growth rate or which are especially rare. It is also feasible for producing "assorted" plants, ie several cultivars grafted onto one root stock, to produce, in the eyes of some people, a plant of increased value. It can also be used to regularise heteromorphosis of branches or buds, effectively creating an improved cultivar.

(1) Root Grafting This can be carried out from the last few days of August until the soil is frozen in winter. The roots of tree or herbaceous peonies can be used as root stock. Since herbaceous peony roots are short and thick, and their xylem is soft, it is an easy method to implement and the survival rate is high. Also, the growth of such grafts is vigorous, so this method is frequently used. A mature herbaceous root tuber without diseases and pests about 25cm long and 1.5 - 2cm in diameter is selected. After being allowed to dry for two to three days, the loss of water makes the tuber become soft and the operation is facilitated.

Healthy annual shoots 5 - 10cm long without diseases and pests are used as scions. If there are not enough such shoots, older twigs can be used but the survival rate is lower and the root growth slower. The scions should be conjoined immediately after cutting.

Notch grafting or cleft grafting is used. A wedge 2 - 3cm long with opposite faces at an acute angle laterally is cut at both sides of an axillary bud at the scion base. Then the top of the root stock tuber is cut off smoothly and a longitudinal cut is made. The length of this cut should be slightly greater than the scion wedge, and it should reach the centre of the tuber, so that it can contain the scion wedge. Both the cut surfaces of the stock and the scion must be flat. The bottom of the scion is inserted in the top of the stock cut and the scion pushed down into the root stock so that the cambium layers exactly touch. The two are tied together with flax and sealed with mud or liquid wax. Then the graft can be planted or heeled in.

Thoroughly dug and manured soil is prepared and levelled. The grafts are planted in rows with the terminal buds of the scion about 3cm above the soil surface after the soil is compacted. Then the grafts are completely buried by loose soil to form a roof-shaped bank for winter protection. Part of the bank is removed the following March, leaving 1 to 2cm of soil for young shoots to come through. They are watered as necessary and weeded. The bank should be maintained so that roots can be produced from the join. In early April flower buds are removed so that all nutrient can be used for young shoot growth. Branches at the top above the bank should be removed from the year-old graft towards the end of September to encourage more branches so that more roots will grow the next year.

In the second year, fertiliser is applied in early March, early May and late August, and the soil is irrigated as necessary. Diseases and insect pests should be controlled. After two years grafts can be transplanted towards the end of September. The herbaceous peony root stock is completely cut off if sufficient root has grown from the scion, but only a half or one-third removed if there are few or even no new scion roots.

(2) Scion Grafting

a. Base Grafting: This is carried out just above ground level. The best time is towards the end of September. Seedling tree peonies are used as root stock and cut across 5cm above ground. The current year's healthy base shoots are selected as scions. An oblique wedge about 3cm long is cut at both sides of an axillary bud at the base of the scion. The stock is split to a depth of about 3cm and the scion is inserted into the root stock so that the cambium of scion and root stock are in contact. The graft is bound firmly with flax or plastic tape. The scion is covered by earthing up for winter protection and other management methods are the same as those for root grafting.

b. Side Grafting: This method is frequently used for grafting at a higher position on a plant to effectively change a cultivar or to cultivate a plant bearing flowers of different colours. It is carried out between mid-July and mid-August. A tree peony is used as the root stock. Healthy basal shoots from selected cultivars are used as scions and one or two leaves are retained. At the back of the scion at a bottom bud, a diagonal cut about 1.5 - 2cm long is made. Then on the other side a wedge 0.3 - 0.5cm long is made with another sloping cut. Next, on the smooth part at the base of a current year's twig with one or two buds, a diagonal cut of 1.5 - 2cm is made to reach one-third to one-half the depth of the branch. The scion is immediately fitted into this cut on the stock with the cambium of stock and scion in contact, and flax string is used to bind them together. Above the graft, one-third to a half of the stock branch should be removed at once in order not to tear the union through excessive weight. When the budded scion is seen to be surviving, any auxiliary buds on the stock are removed to ensure maximum nutrient supply for the scion. The string bound around the cuts is not released until the two parts are firmly united. Then the remaining part of the stock and any bottom buds are removed and some additional manure is spread to help growth. While the side graft is becoming established on the stock branch, it is best to keep the stock plant shaded to prevent excessive transpiration.

(3) Bud Grafting Bud grafting can be carried out between April and August when the bark and phloem of branches can be peeled off. However, the survival rate is greatest if it is done during May, June and early July. All tree peonies can be used as stock. Plump buds from the current year's growth are selected as scion material. Buds from 2- to 3-year-old branches can be used if budding is done during April and May.

Two cuts are made 1.2 - 1.5cm above and below the selected bud. The width of the bud piece is generally half that of the scion with minimum width at least 0.5cm. The bud is taken off with part of the xylem, and any leaves are immediately removed. Then a cut slightly smaller than the bud piece is made at an axillary bud at the grafting location on the stock. The bud piece is gently peeled with a grafting knife leaving the wooden bud knob. Then the xylem of the stock is lifted with the tip of the knife and the bud piece with the bud knob is swiftly embedded into the stock cut. The scion bark around the bud piece is trimmed if necessary and plastic tape is used to seal the cuts tightly, with only the bud piece and petiole left outside.

If a bud piece is grafted onto a 2- to 3-year-old branch, a cut the size of the bud piece is made in the stock and the stock bark is lifted up. The bud piece is quickly put into the stock cut and then is tightly tied up. All leaves, with their petioles, on the stock above the bud piece are removed but those below the bud piece are retained. The success of the graft can be determined 10 - 15 days after grafting. If the grafted bud is fresh and swelling, it is alive, and excessively tight binding should be loosened. If the graft has failed, a replacement should be made. Before sprouting in the following year, the stock 1cm above the bud is cut off after the bud begins to develop. Basal shoots are removed.

3. Propagation from Seed

This is mainly used for large scale production of root stock for grafting; also of course to obtain new cultivars.

Tree peony seeds begin to ripen in late August. The complete seed heads can be harvested when their skins turn brownish-yellow. The seed heads are spread out to mature and dry in a cool and well-ventilated place, possibly indoors. When the seed skins become black and the follicles crack naturally, the seeds can be removed, and after another two or three days for drying, they can be sown. Sowing is best begun in early September for a high germination rate. To stimulate early germination, the seeds are placed in wet sand. After two months young roots are produced and the seeds are then sown in earth. If seeds are very dry at the time of sowing, they can be soaked in warm water for 24 hours. Soaking in sulphuric acid for 2 to 3 minutes, or in 95% ethyl alcohol for 30 minutes, are also helpful.

Sandy soil is chosen for seedling cultivation, and free-draining and approximately neutral soil is recommended; also it should be well fertilised. According to experience in Heze, 30 - 45 tonnes of manure or 6 - 7.5 tonnes of cake fertiliser per hectare is appropriate. If the soil is very dry, irrigation and the addition of organic matter for water retention is needed before deep digging and leveling.

Sowing in rows is usual but seed is also broadcast. About 150 kg of seed is used per hectare, and covered with 2 to 3cm of fine moist soil.

Tree peony seeds revert to a dormant state after germination, and only roots develop in the first year. If seeds are soaked in 0.6 - 1.0 mL per litre gibberellin for 24 hours before sowing, or if after root development gibberellin is watered in once or twice per day, the dormancy can be broken after one week.

4. Layering

(1) Ground Layering This method is useful for varieties with few basal shoots or a less developed root system, but the rate of production of new plants is low.

Layering is done at the end of May and beginning of June after flowering. A strong and healthy two- to three-year-old branch is selected and pressed down. It is notched at the junction of the current year's growth and the older part of the branch. It is laid in the earth and fixed, usually with weights. The soil should be kept moist for new roots to germinate. It is more beneficial for new root development if the part of the old branch not pressed into the earth is also notched nearly to breaking but still connected. When sufficient fibrous roots grow, the rooted branch can be cut away from its parent tree and it becomes an established new plant before the next winter.

(2) Air Layering The success rate of air layering is highest when it is carried out ten days after flowering, when the branch is half lignified. A girdling cut with a width of 1.5cm is made about 1cm below the second or third leaf axils at the base of the twig. Then absorbent cotton dipped in 0.5 - 0.7mL per 10 litres indolebutyric acid or 0.4 - 0.6mL per 10 litres solution of hormone rooting powder is placed and secured around the root producing area above the upper part of the girdle.

Plastic film is tied round this and then cut to form a cylinder. A wet and loose mixture of ground slag and moss mixed with clear water is fed into the cylinder, which is then sealed and supported by poles. Immediately after the "bag" is suspended, 30mL of clean water is injected, and then 30 - 50mL of water is given every 15 - 20 days. When roots germinate, after one-and-a-half to two months, liquid nutrient is supplied instead of water. When the new young roots can be seen, the branch can be cut from the parent plant and transplanted into a pot or a nursery bed for further cultivation and care. The success rate can reach 70% or more.

5. Cuttings

Although peonies can be propagated by cuttings, the slow growth and difficulty of maintaining the precise care needed mean that this method is rarely adopted for production. Cuttings are taken in September. Current year's growth of 15 - 20cm long with 2 - 3 buds is used for cuttings. On planting cuttings, shading and control of moisture are needed. Roots can appear after 20 - 30 days. Cuttings can be treated with 0.5mL per litre naphthaleneacetic acid, or 0.3mL per litre indolebutyric acid, or 0.5 - 1.0mL per litre gibberellin, to stimulate root production.

In order to increase the survival rate of planted cuttings, ridging can be carried out in late autumn or spring.

6. Double Levelling Method

This is a new technique for rapid peony production. The advantages include convenience and ease, saving of labour and time, high production rate, and rapid new plant growth in large quantities and at a uniform pace. It can also be applied widely.

This method of propagation can be used at any time during the dormancy period, but is best done in September and October. A parent plant is dug up and divided immediately into smaller plants, each with one or two branches. Then these plants are planted in holes of 30 - 35cm in depth, 25 - 30cm in diameter and 70cm apart in prepared ground. On one side of the holes, the location for level underground layering, a hole 20cm wide, 10cm deep and the same length as the branches is excavated for the branches to be pressed into the soil when two-thirds of the base of the plant is covered. The tips of the branches should be lower than or at the earth surface while the lower part of the branch is buried slightly deeper. When covered by soil, perhaps in ridges, and compacted, the buried plants are thoroughly watered. Manure should be

dug in before planting, and top-dressing fertilisers should be heavily applied during growth. If the apical buds grow normally in the first spring after branch burying, the developed branches can be cut during September and October. Because the apical dominance is eliminated when the branches are buried, many lateral buds develop in the next spring. The new branches that develop are pressed into the ground in September or October of the same year so that even more shoots come from the buried branches. In the third autumn the whole plant is dug out and divided into several new small plants according to the root development from the buried branches and the branches above the ground. Generally, a parent buried branch can form eight to ten new young plants.

Cultivation

Chinese Mudan have a wide ecological adaptability so that their cultivation is easy and conveniently managed. They can be grown both in gardens and in pots in all parts of China with the exception of Hainan, the southernmost province.

1. Selection of Appropriate Cultivars and Sites

The four cultivar groups of Chinese tree peonies have ecological characteris-tics and soil conditions that suit them best. Before planting, suitable sites or the properties of a cultivar group and its individual varieties should be considered for future cultivation and management to ensure success.

Chinese peonies are classified as temperate plants. They tend to thrive in cool, dry and sunny places. As they have prominent long fleshy roots, the best choice for their cultivation is a dry and airy site on higher ground, with lateral shading, good drainage and deep porous fertile soil. Uncultivated or sticky or saline earth and poor drainage are best avoided. But different cultivar groups, and cultivars in a group, can to some extent be suitably matched with the ecological environment. In addition, land that has been cropped continuously with peonies should be replanted with peonies after one to two years rotation. After the site has been selected, a generous quantity of manure should be applied and deep digging and leveling should be repeated several months or half a year before planting. The ideal soil for planting in pots is also loose, fertile, well-aerated and well-drained. Future developments in pot cultivation are likely to involve artificial nutrients and an inert medium such as perlite.

2. Planting

Autumn is the best season for planting peonies, because it helps early development of new roots and thus early recovery of plant vitality for normal growth in the next spring. If a peony is divided and planted in spring, its root system cannot recover before the onset of rapid growth and development of the parts above ground. Thus the nutrient supply cannot match the need and the plant may greatly lose its vitality, and as a result normal growth and flowering will not return until after several years' cultivation. Planting should take place during the period from mid September to the end of October. The earlier part of this period is better, because the soil temperature is still high then and this benefits the early growth of new roots from divided plants, which ensures their winter survival and development the next year. Excessively early planting may result in "autumn bud-burst" - shoots that develop into branches too soon and which will be damaged by cold during winter, because they have not lignified. Very late planting is disadvantageous to new root development which results in slower recovery and a lower survival rate.

If possible, planting should be carried out promptly after division. The planting space should be broad and deep. Well-rotted manure dug in with the bottom soil will encourage the root system to distribute evenly. The correct planting depth is with the collar the same level as or slightly below the earth surface. Soil should be trodden in layer by layer. Finally the plants are thoroughly watered.

3. Irrigation and Feeding

Chinese peonies consume a great amount of nutrient and also transpire a great deal because of their luxuriant branches and leaves and huge flowers, and although they thrive in a dry environment they do not grow well without moisture. The principle of watering is to keep soil moist but not too wet and definitely not flooded. Generally, newly planted seedlings should be thoroughly watered, while watering of established plants should be related to the growing environment and individual flower conditions. In northwest China, for example, there is less annual rainfall than elsewhere. Less rain in the dry spring and less rain and snow in the long winter mean that the supply of an appropriate amount of water and attention to soil moisture conservation are particularly important there.

The application of fertiliser in proper amounts and at the proper times not only facilitates plentiful flowers of large size, bright colours and full flower forms, but also prevents or reduces the differences between on-year and off-year flowering and the degeneration of flower forms. Ideally, fertiliser is applied three times per year. Flower fertiliser, organic fertiliser with nitrogen and phosphorus, when soil thaws and shoots germinate aids the growth of branches and leaves and the rapid development of flower buds. This application of fertiliser is half the total yearly quantity. Shoot fertiliser, a rapid-effect compound fertiliser, is applied after flowers fade. It serves to replenish plant nutrient already consumed to maintain growth and to promote smooth new flower bud differentiation. The third application, a surface mulch to preserve soil moisture, is applied with irrigation before winter and freezing begin. The use of decomposed manure or dung at this time as the mulch ensures a safe winter and provides nutrients for the next spring. At any other time, except high summer, light liquid compound fertiliser application with watering will help better plant development if necessary.

4. Pruning

Pruning or trimming affects plant growth, flower quality, visual effect and life span. The timely elimination of dry, sick or pest-damaged branches and of redundant branches and buds can maintain the dynamic balance between the parts of the plant above ground and below ground, and the even distribution of branches. It also ensures good ventilation and illumination.

(1) Selection and Maintenance of Stems and Branches After peonies are planted out, many new shoots appear at the collar in the first year of growth. When they develop to 10cm in the second year, those that grow healthy and are evenly distributed are kept to provide new stems and branches, and the others removed, to ensure an elegant shape. In later years, one or two new shoots are left every one or two years so that the plant becomes larger and gradually fuller as desired.

(2) Utilisation of New Shoots To develop flowers of huge size and bright colour, bud thinning should be carried out at the same time as pruning. Only one healthy and strong shoot is kept on each branch. Adventitious buds on old branches should be completely removed to concentrate nutrient for the flowers. On vigorous varieties more buds can be retained to increase the number of flowers and to prolong the flowering period. In addition, faded flowers where the seed is not required should be removed. The dried tips of branches should be removed before winter or in early spring.

5. Disease and Pest Control

(1) Peony Grey Mould (Botrytis) This is a serious fungal disease of peonies in all countries and can occur at any time. Stems, leaves and flowers can be affected. When a young plant is infected, wet brown scabs and rot are seen at the base, then growth stops and the plant withers. When infection occurs on leaves, wet-looking spots appear at the edges. Most are brown, some purplish-brown, occasionally with irregular wheel patterns. A film of grey mould is produced on sick areas in wet weather. On petioles and stems, scabs are always dark brown strips and slightly sunken. The diseased part usually rots. Infected flower buds turn brown and dry, and rot. Low temperature and humidity in June and July, dense clusters of foliage and flowers, and excessive nitrogen fertiliser encourage the disease.

Prevention and Control: **a.** Sick and withered branches and fallen leaves in autumn, diseased buds and leaves in spring, are removed and deeply buried or burnt. **b.** Spraying two or three times, at 10 - 15 day intervals, with 1% Bordeaux mixture with equal amounts of lime and chalcanthite, or 70% thiophanate methyl solution 1 000 times diluted, or 65% zineb solution 500 times diluted, or 50% botran 1 000 times diluted, is an appropriate treatment. **c.** Preventive measures include proper planting density, good drainage, crop rotation in seriously affected areas, and soaking young plants for 10 - 15 minutes in a fungicide before planting.

(2) Peony Brown Blotch (or Red Blotch) This is also a fungal disease which occurs in all countries. It affects the leaves, sometimes branches, floral organs and seeds. At the onset, small brown nearly circular spots appear on leaves, and expand into circular or irregular large brown spots on both sides. In serious cases, the blotches merge and the leaves become withered and twisted. In wet conditions, the backs of the blotches are covered with a dark-green film of mould. On infected stems or petioles, the infection appears in strips. The purplish-brown blotches are initially slightly convex and then concave, and later the centre splits. When the disease takes hold at the base of a branch, the branch is liable to break. Small brown spots can be seen on flower organs, and the edges of petals appear burnt, when the infection is heavy. This disease reaches its height from May to July, particularly in a warm and rainy summer. Not removing sick and damaged plants, excessive application of nitrogen, high planting density, poor ventilation and excessive exposure to sunshine encourage the disease.

Prevention and Control: **a.** Removal of infected parts, and thorough removal of infected plants in autumn. Before growth begins in early spring, spray with lime sulphur or 50% carbendazim 600 times diluted. **b.** During the growing season, spraying 3 or 4 times, at 7 - 10 day intervals, with 50% carbendazim diluted 1 000 times, or 65% zineb diluted 500 times, is an appropriate treatment.

(3) Peony Anthracnose This disease damages leaves, stems and floral organs of a peony. In the area around Beijing, small brown spots appearing in June expend gradually into approximately circular scabs with a diameter between 4-25mm in accordance with different varieties. Along leaf edges the scabs are shaped like semi-oval. Because of the blockage from the main leaf veins, the expanded scabs slways show a semi-oval shape as well. The colour is blackish-brown and then become gray white in the center and reddish brown around the rim. Many tiny black spots scattered in the scabs. In moist condition reddish brown gloeospore groups which is the typically features of anthracnose pro-duced. During the later period the sick scabs split and perforate. While long scabs with a diamond shape of 3-6mm can be seen from stems and petioles. They show reddish-brown, but grayish-brown with a reddish-brown fringe at last. Sick stems are often twisted and even broken. Twiggeries in morbidity die very quickly. If a perula or a petal is infected, the bud will wither up soon and the corolla will deform. The disease is still caused by mycotic infection. The infection appears in June and reaches its height from August to September in Beijing area. Heat, moist and high plant density encourage the disease.

Preventions and controls: **a.** Removal of infected parts (see Peony brown blotch). **b.** During the growing season, spraying 2 or 3 times, at 10-15 day intervats, with 70% paroxysm,1% Bordeaux mixture with 1:1 lime and chalcanthite,or 65% zineb diluted 500 times as soon as infection appears.

(4) Peony Zonate Spot or White Star Disease This fungal disease mainly damages peony leaves, and causes them to wither. Tiny light yellow spots appear on leaves and gradually become larger, approximately circular, 3 - 10mm in diameter, and light to mid-brown. Later, the centre of the spots turns brownish-grey and there are very obvious circular brown scabs on either side of the leaves, with fine mildew concentrated at their centre. At leaf edges and near main veins the spots are semi-circular. In serious cases, 20 - 30 spots may be present on one leaf.

In Beijing the disease starts in June and reaches its height in August and September. It is mainly a disease of older peonies, and occurs when plants are unhealthy, or there is excessive rain and moisture, or too great planting density, or bad drainage and waterlogging, or a lack of good ventilation and sunlight.

Prevention and Control: **a.** Remove infected and fallen leaves and spray with 3% lime sulphur during dormancy. **b.** During the growing season, spray two or three times at 10 - 15 day intervals with 1% Bordeaux mixture, or 50% Tuzet diluted 800 times, or 65% zineb diluted 500 times. **c.** Preventive measures include proper planting density and good ventilation and sunlight.

(5) Peony Leaf Blight This is another fungal disease known in China and also reported in the former Soviet Union and Japan. It infects only leaves and causes early leaf fall. Initially, tiny brown spots appear on leaves. They expand to form circular scabs 1 - 3mm in diameter which are yellowish-brown and brown and slightly depressed on the surface, but dark brown at the back with a red fringe. Black grains, the pycnidia of the organism, are scattered on the scabs. Later the scabs drop leaving a perforation. Prevention and control are the same as for brown blotch.

(6) Peony Branch-Rot This is a fungal infection that infects stems and branches. The symptoms are light brown scabs on stems, later becoming oval and reddish-brown. When they completely encircle a stem, the stem and branches above will soon die. In autumn, black grains, the pycnidia of the organism, appear on the scabs. Infected buds turn brown and remain on the plant where they rot and die. Pathogens infect mainly through wounds. Cuts and weak plant growth facilitate the disease.

Prevention and Control: Strong growth and few wounds improve resistance to the disease. Chemical treatment is the same as for Peony Zonate Spot.

(7) Peony Root-Knob Nematodes This disease, caused by northern root-knob nematodes, is known in China and other countries. In the last ten years in China this has been the most serious disease to affect root systems. It causes leaves to fall early.

The root-knob nematodes only damage nutritive roots. The first sign of the disease generally appears after flowering when the leaves turn yellow at the edges, wither and fall early. Plant growth is stunted if the infection continues for several years. Meanwhile on nutritive peony roots knobs of different sizes form. The young knobs are yellowish-white, hard and tough, and 2 - 3mm in diameter. Later on, the epidermis tissues on the knob break. When cut open, shiny white spots can be seen. The disease is characterised by 200 - 300g dense clusters of fibrous roots with knobs developing on them. The eggs and larvae of the nematodes spend the winter in knobs on the roots, or in the soil, or on wild host plants. In the spring the larvae directly infect newly developed nutritive peony roots. The nematodes are spread by movement of soil, flowing water, on cultivation tools and by transplanting seedlings.

The periods when the infection is at its height in the Beijing area, ie May to June and October, correspond to the two dominant periods of nutritive root growth. There is evidence of a relationship between the disease and certain peony varieties. Those varieties with their roots mainly distributed to a depth of 15cm are the most vulnerable. The nematodes enjoy a wide range of hosts, and also damage many other plants.

Prevention and Control: **a.** Attention to plant quarantine to prevent the expansion of infected areas. Treatment of infected seedlings consists of soaking for 30 minutes either in 0.1% formaldehyde or in water at 48 - 49°C. **b.** A chemical preventive measure in the field is the insertion of granules of 15% aldicarb at a depth of about 10cm annually (at the beginning of May in Beijing). **c.** Infected soil is treated by drying (0.17% moisture) and high temperature sterilisation, or by pesticides. These include methyl bromide, aldicarb and fenamiphos.

Peonies And Their Uses

Mudan, with their large flowers, splendid colours and impressive looks, are extremely valuable as ornamental plants, so they have been widely used to add beauty to landscapes both in small groups and in mass planting. They are also of significant economic value.

Peonies in Landscape
Peonies are extensively used in various types of urban planting, for instance in parks, gardens, street green belts, temples and classical landscapes.

1. Specialised Peony Gardens
In some specialised gardens, deliberately constructed as a whole or as part of a larger scheme, peonies are the subject and form the main part. A great number of excellent peony cultivars are planted together to enhance appreciation of the richness and variety of their shapes, colours, fragrance and charm. Either a formal or a natural arrangement is commonly adopted.

(1) Formal Arrangement Also called Geometric Arrangement, this approach is generally applied in comparatively flat topography. Regular flower beds, in which the peony plants are evenly spaced, are designed in the garden area, resulting in an orderly geometric pattern. Since other plants or rocks are rarely included, this has the advantages of easy cultivation and administration, maximum acce-ssibility for visitors, convenient comparison, study and apprecia-tion of peony cultivars. On the other hand, this arrangement does not provide an environ-ment that serves as a foil to the peonies, so they are unable to play their full role in the landscape.

(2) Natural Arrangement This approach, also called Scenic Arrangement, refers to the harmonious integration of peonies and other flowers with trees, rocks, hills and buildings, to create an effect of artistic naturalness, as if not man-made. The inherent beauty and distinction of the peonies are shown to their advantage in an attractive and contrasting setting. One can stroll along a winding hill path through constant changes of scenery and become so enchanted as to forget about going home. Especially for those who have spent a considerable part of their lives in urban areas, a feeling of "going back to Nature" will involuntarily well up in their hearts. The natural arrangement is the best in which to appreciate the growth of peonies, so that the finer properties of the cultivars can be highlighted and their ornamental effects fully displayed. Since most wild peonies grow in open woodland, peonies benefit from being interspersed with other plants which provide shelter from sun light. A further advantage is that the gardens do not appear dull after the peony flowers have faded if other flowers and trees are also planted. Most of the municipal peony gardens incorporate artistic creations that further stress the theme of the garden. The aesthetic charm and subjective features of peony gardens are emphasised by, for example, the peony tower and statue of the Peony Fairy Maiden in Wangcheng Park, the statue of the Peony Fairy Maiden in Caozhou, the peony gallery in Shanghai Botanical Garden, the peony pavilion in the Flowery Waters in Hangzhou, and the statue of the sleeping Peony Fairy Maiden in Beijing Botanical Garden. In the last of these, a gateway vividly describes the inextricable and commiserative love between the purple peony Gejin, the white peony Yu Ban, and two Luoyang gentlemen. It is pleasant for visitors to be both intoxicated with the unique beauty of the peonies and immersed in the artistic conception of peony legend and mythology, enhancing their enjoyment, thoughts and emotions. A wall in Xiyuan Park of Luoyang, made of tri-coloured glazed pottery initi-ated in Tang, expresses the grand occasion when Wu Zetian, escorted by a big group of palace maids, went to the Imperial Garden to enjoy the peonies. Such artistic creations show that the graceful and glorious peony is able to combine appropriately with classical constructions, the buildings in ancient style providing the peonies with an attractive background and environment. The peonies, in their turn, enhance the splendour and vigour of the ancient artistic works. When a peony garden is designed, supplementary planting of Shaoyao greatly prolongs the ornamental period because the herbaceous peonies come into flower as the Mudan flowers are fading.

2. Peony Terraces
Peonies are best planted in high and dry places, avoiding dips and depressions, so in areas of low-lying land or of high water level because of more rainfall, terraces should be constructed. Each flower terrace, or raised bed, should be 60-100cm above the ground level. Those built on a slope can be in the style of steps – this tier upon tier effect is better than a single-layered terrace. For terrace planting, the cultivar shapes, heights and flower colours should be carefully studied to achieve an appropriate grouping and an aesthetically pleasing eleva-tional result. Terraces are a good way to bring visitors closer to the peonies. Formal terraces are built with bricks, stones and concrete in regular geometric shapes and the peonies are equally spaced. Natural terraces are generally constructed with mountain stones and rocks, following uneven topography to

form irregular shapes and curves. This gives the effect of a natural wild and mountainous environment full of interesting variation. Within the scene, exquisite rocks and other flowers, shrubs, grasses and trees act as a foil to the peonies.

3. Peony Belts

Peony belts are usually constructed along both sides of a path in parks or gardens, or in the central reservation of main roads in urban areas. The use of Mudan is an important component in the spring scenery. During the peony flowering season, people can enjoy their pretty shapes, blazing colours and fragrance as they stroll along a garden path or drive down an urban street. For instance, in Luoyang, the central dividing strip in Zhongzhou Great Street is planted with large numbers of mudan, interspersed with Shaoyao, deodar, myrtle and Chinese roses. Thus there are flowers for three seasons, while there is greenery throughout the year. This has become one of the most charming street scenes in Luoyang, and its example has initiated the use of peonies in a prominent role in street plant-ing projects.

4. Group Planting and Mass Planting

Planting in groups and mass planting of peonies is often carried out at the edges of woods, in lawns and beside rocks. Mass planting emphasises that mudan is an ideal plant for creating a flowery environment in a simple and natural style.

Peonies in Interior Decoration

1. Peonies in Pots

During the Spring Festival, families in Guangdong province as well as in Hong Kong and Macao buy flowering peonies in pots. They are a celebration of the festival and also an ornamental symbol of people's expectation of wealth and luck in the coming year. Since the last years of Ming, it has become an important part of the peony business in Heze and Luoyang to meet the demand for flowers at Spring Festival by forcing culture. Potted peonies are an attractive element in room decoration.

2. Peonies in Flower Arranging

Mudan play an essential part in the Chinese traditional art of flower arranging. As early as Tang, peonies had become important as cut flowers with a highly honoured place in floral arrangements in the imperial palace. The process of producing and appreciating artistic works with flowers was strictly regulated. The huge size, brilliant colours and striking ornamental effects of Mudan flowers makes them appropriate for inclusion in middle and large scale arrangements in halls, meeting rooms, studies and sitting rooms. Further research on the production of peonies as cut flowers is needed, to develop their use in flower arranging.

3. Bonsai Peonies

The use of peonies as subjects for bonsai is a new development. When the splendid peonies of "National Beauty and Heavenly Fragrance" are integrated with either unsophisticated or elegant examples of the art of miniature landscape, the new effects are highly appreciated. It is therefore suggested

Peony belts

Peonies in flower arranging

that this use of mudan for interior decoration and enjoyment of the flowers is greatly extended.

Peony Production

1. Production of Young Peony Plants

The scale of production of peony seedlings and young plants continues to increase. Besides the gradual enlarging of the domestic market, the growth of annual exports of plants to Japan, North America and Europe is evident. To suit international markets, intensive production is needed of large batches with uniformity and systematisation of various cultivars (early, middle and late flowers of all colours).

2. Production of Peonies in Pots

The production of peonies in pots brings great economic benefits. The development is towards lighter and smaller plants, and also soil-less culture. The cultural technique of forcing peonies has tended to be used to supply plants in pots to Guangdong, Hong Kong and Macao for their Spring Festivals. However, Mudan in pots are probably now produced more widely for the Chinese societies in southeast Asia and Japan. For the domestic market, smaller and lighter potted plants are produced for sale during the flowering period.

3. Production of Mudan for Cut Flowers

Fresh cut peonies are an important flower material for both domestic and international markets, with a large annual consumption worldwide. Until now Japan has been the centre of production. Improvements must be made in China in selecting and cultivating Mudan varieties for cut flowers, and in cutting techniques, storage, packaging, transportation and marketing.

Bonsai peonies

4. Production of Peony Bonsai

Bonsai of peonies offers a contrasting approach to appreciation of the aesthetic beauty and charm of the flowers. Although the stems of peonies are hard, a large part of their annual shoot growth withers during over-wintering, which results in comparatively small annual stem production. Topiary work every year gives the plants a new look. Bonsai peonies, with their huge flowers, gorgeous colours, attractive fragrance, dancing leaves and beautiful shapes at flowering time, are rewarding for both appreciation and economic benefit. Further study is needed to standardise and improve the production process.

5. Mudan in Medicine and Other Fields

The earliest recorded use of Mudan in medicine can be found in *Shen Nong's Herbal Classic*. The part of the plant for medicinal use is the bark of the root system, called Danpi or cortex Mudan. Danpi contains several medical components, the important one being paeonoside ($C_9H_{10}O_3$). It functions as an antibiotic to diminish inflam-mation, as a tranquilliser, analgesic, antispastic and hypotensor. Since it is an important Chinese herbal medicine, a great quantity is produced every year to meet demand in China and abroad. It takes five years to complete the production of Danpi, from sowing, transplanting seedlings to fields, and digging up roots, but only three years are needed if plant divisions are used. The process of making Danpi includes scraping the epidermis and removing xylem. If the epidermis is not separated, the product is called "original Danpi". After Danpi is dried in air and graded for packing, it is sold in markets.

Mudan are also edible and taste delicious. If used in wine fermentation, the strong scent of the flower when the wine is drunk quietly and exquisitely gladdens the heart and leaves endless aftertastes. In addition, essence can be extracted from the flower, and there is great potential in the cosmetic and health food industries because peony pollen contains various nutritional constituents.

FAMOUS PEONY GARDENS

The cradles of production and development of Chinese Peonies are the Yellow River and the Central Plains where they have become a beautiful and unique group of plants.

The love of peony flowers is a strong tie linking all the descendants of the Chinese nation, and the peony flower has become a symbol of the nation's pride. The cultiva-tion of peonies pervades all parts of the country except Hainan Province, and famous peony gardens can be seen all over China. The peony flowering season is one of the nation's great occasions, when all manner of people pour into gardens to appreciate and enjoy the sumptuous blossom.

Some famous peony gardens are briefly introduced here.

1. The Peony Garden in the King's Town (Wangcheng) Park, Henan

The name comes from the location, which is the site of the ruins of the King's Town in ancient East Zhou Dynasty. Constructed initially in 1956, this garden has become an important place in Luoyang to enjoy penies. In its total area of $37.07hm^2$, 19800 peonies of 320 varieties are planted in $13\,340m^2$. Over 20 varieties have been introduced from Japan. Some delightful entertainment places have been built in the park, such as a peony pavilion as well as a parterre group with a Peony Maiden sculpture. Between April 15th and 25th, the yearly peony fair attracts many thousands of people, reaching a peak of 250 000 people daily.

2. The Peony Garden in the Peony Park, Henan

This park in Luoyang is well-known for its peonies. It was built in 1956 and covers $5.87hm^2$. Peonies are the main flower in the park, occupying $5660m^2$. In the 10 peony parterres 3 960 peonies are grown, including over 200 varieties.

3. The Peony Garden in Xiyuan Park, Henan

Originally named Botanical Garden, it was renamed Xiyuan (Western Landscape) because the location was once the western landscape of the Sui Dynasty. Built in 1958 with a total area of $13.14hm^2$, over 6000 peonies from nearly 200 varieties are planted in over $10\,000m^2$.

4. Guose (National Colours) Peony Garden, Henan

Since it was built in 1985, 200 000 peonies in about 400 varieties have been planted in $10hm^2$. In 1992 it was designated the State Peony Gene Pool by the Forestry Ministry of the People's Republic of China. Since it is on Mount Mangshan and the flowering time is later than in the city, later visitors to Luoyang can still enjoy the beauty and fragrance of the flowers here.

Wangcheng Park, Henan

Peony Garden in Xiyuan Park, Luoyang

Peony Garden in the Peony Park, Luoyang

Zhaolou Peony Garden, Shandong

5. Luoyang Peony Garden, Henan

Built in 1992, it covers a total area of 10hm^2. In its 6.7hm^2 380 varieties amounting to approximately 100000 peonies are cultivated. In 1994, more than 200 plants of 11 famous foreign varieties, including the Japanese 'Jin Di' and 'Tai Yang', were introduced.

6. Caozhou Peony Gardens, Shandong

Located in the Peony County, Heze of Shandong province, they consist of three gardens, Zhaolou, Liji and Helou. In the total area of 73hm^2, there are over 400 peony varieties. The Caozhou Peony Gardens are the largest peony gardens in the country with the greatest number of varieties. Since the peony is the city flower of Heze, the period between April 22nd and 28th each year is entitled Heze International Peony Fair. Millions of domestic and foreign visitors go to see the flowers because of their reputation.

(1) Zhaolou Peony Garden The entrance to the 24hm^2 Zhaolou Peony Garden is an antique elegant and colourful *Paifang* (memorial archway) with the Chinese words Caozhou Peony Gardens on it. There are over 400 tree peony varieties and over 200 herbaceous peony varieties. An interesting and novel building of classic style called the Watching Flowers Tower is built in the garden. In front of it, two erect stone tablets are inscribed with Pujie's The Greatest Fragrance Under the Heaven and Shu Tong's Caozhou Peonies over the Top of the World. Selection and breeding of new cultivars has been carried out for many years in the garden. As a result, some well-known varieties such as 'Fen Zhong Guan', 'Guan shi Mo Yu' and 'Zi Yao Tai' have been cultivated.

(2) Liji Peony Garden The garden, located to the north of Zhaolou Peony Garden, also has a colourful *Paifang* as its gate, with the same words Caozhou Peony Gardens on it. In the total area of 20hm^2 tree peonies are widely planted along with herbaceous peonies and other flowering shrubs. By climbing the antique four-storey Tower of Heavenly Fragrance, the panorama of the garden can be seen. A huge banksia rose with a curved and twisted trunk gives off a rare fragrance, adding to the feeling of elegance and grace.

(3) Helou Peony Garden It is situated to the northeast of the Zhaolou Peony Garden and covers 26.7hm^2. It is mainly planted with traditional tree peony varieties together with herbaceous peonies and many flowering shrubs.

7. Caozhou Hundred-Flower Garden, Shandong

It was originally built in 1958 to the north of Hongmiao Village of Peony County and called Hongmiao Garden. The government granted funds in 1982 for rebuiling and it was renamed Hundred-Flower Garden. In the area of approximately 6.7hm^2, there is novel and unique gate, a reception building and a pebble-mosaic tourist path. Among 60 000 peonies of over 400 varieties, there are both prestigious traditional varieties and new plants of excellent characteristics. These include 'Chun Hong Jiao Yan', 'Jing Yu', 'Yin Hong Qiu' and 'Bai Yuan Fen'. Also some plants have been introduced from Japan. Many groups of tourists from China and other countries are attracted by the great number of varieties with their own special features concentrated in a comparatively small place.

8. Past-Present (Gujin) Peony Garden, Shandong

Located in Wangli Village, in Peony Country of Heze, it occupies about 1.7hm^2. As flower cultivation has been a tradition in Wangli Village from ancient times, tree peonies here are mainly planted with other flowering bushes. A special feature is juniper weaving and modelling, a traditional art among the local villagers who create the shapes of gate towers, *Paifang*, lions and tigers, etc. A large example is the juniper *Paifang* Diligent Reading *Paifang* (Zisong Fang). This has lasted for over 200 years, and still remains luxuriant and vigorous with a clear outline. In the

Hundred-Flower Garden, Heze

Peony Garden in Jingshan Park, Beijing

Peony Garden in Jingshan Park, Beijing

Peony Garden in Beijing Botanical Garden

Peony Garden in Beijing Botanical Garden

garden, there is also a 100-year-old Chinese pagoda tree.

9. Peony Garden in Beijing Botanical Gardens

The Peony Garden is one of the specialised gardens in the Beijing Botanical Gardens. It was initially built in 1981 and opened to the public in April 1983. In its 6.7hm^2 over 5 500 tree peonies of 230 varieties introduced from Heze, Luoyang and Tianshui are planted. There are also more than 200 herbaceous peony varieties numbering up to 2 500 plants. On a hill, 93 kinds of trees and shrubs are deliberately and skilfully arranged in a natural style. The varied layers and careful distribution have created a splendid environment for growing peonies and enjoyable areas for visitors. Three buildings in the garden, Squared Pavilion, Doubled Pavilion and Tower of A Hundred Flowers, are added attractions. A huge porcelain wall facing the gate describes the myth of peony goddesses Gejin and Yuban. A group of artistic stones in front of the middle gate are carved with four Chinese characters written by Mr Wu Zuoren, a famous painter, meaning "thousand clusters like glowing snow", a vivid depiction of the peony scene in the flowering season. In the centre of the garden, a sculpture of the sleeping peony maiden adds the finishing touch to the whole work of the garden.

10. Peony Garden in Jingshan Park, Beijing

Located inside the East Gate of the Jingshan Park in Beijing, it occupies about 1 200m^2, in which are planted approxi-mately 700 peonies of 130 famous varieties all introduced from Heze in Shandong province. It is the best place to appreciate peonies in downtown Beijing. Since most of the peony plants are several decades old, giant flowers grow on twisted stems and old twigs, showing the typical plant form of the different varieties. The garden, with its regular flower beds surrounded by paths, is divided into two parts. The eastern part contains mature peonies of several decades old, while in the western part, younger peonies are being nursed. The beds are edged with low hedges of Chinese trees which provide protection and an excellent background for the peonies. The careful administration and cultivation of the large tree peonies is based on a rich accumulation of experience. The brilliant peonies of various famous varieties grouped together display to perfection a blaze of colour.

11. Peony Garden in Zhongshan Park, Beijing

In the front of this park, to the west of Beijing's Tiananmen Square, 1 000 plants of over 100 well-known varieties are cultivated among ancient cypresses. The ten-year-old large-flowered plants grow luxuriantly and clearly benefit from careful administration. Tourists deeply appreciate the inspirational sight of the brilliant flowers in an open woodland setting.

12. Peony Garden in the Imperial Garden of the Forbidden City, Beijing

Tree peonies have been widely planted in the Imperial Garden since the Qing Dynasty because the tree peony was then the state flower. One feature of the garden is the way that the dignified and graceful peonies and the resplendent and magnificent ancient buildings contrast with and complement each other. Another is the natural and harmonious arrangement of flowers planted in parterres or with mountainous rocks and sword-shaped stones.

13. Peace Peony Garden, Gansu

This garden, started in 1967, is in the southeast suburb of Lanzhou in Yuzhong County of Gansu Province. It contains over 200 varieties. The Central Plains varieties have also been introduced from Luoyang and Heze. The breeding work has involved cultivated plants of authentic ancestral species, particularly *P. rockii*, which were either collected from the wild or introduced from other sources. It is the largest cultivation and reproduction base of the Northwest peony cultivar group in a peony-growing area of 8.7 hm^2, and includes 130 000 plants, many of Gansu Mudan type. Some famous new varieties, such as 'Hei Tian E', which flowers in May.

14. Peony Garden in Mount Danjingshan, Sichuan

It is located in Jiulong Town, 16km away from Pengzhou city, Sichuan province. This city was named Tianpeng in ancient times and was famous for peonies as early as the Dynasties of Tang and Song. In 1985, the tree peony was adopted as the city flower. Peonies have been widely planted and can be appreciated in several areas. These include Peony Ground, Heavenly Fragrance Garden, Red Sun Glow Garden, Eternal Peace Yard and Steles Forest Garden on Mount Danjingshan with an elevation of 1 147m. They are the gardens in the southwest part of China where the greatest quantity of tree peonies are planted, numbering about 100 000 peonies of over 200 varieties. In addition, a peony garden in downtown Pengzhou Gardens occupies 0.7hm^2 and more than ten old local varieties are cultivated. The yearly peony fair attracts tremendous numbers of people. When the flowers fade in the downtown areas, those on Mount Danjingshan are still in bloom. Because of their later flowering, the total period for people to enjoy peony flowers is increased.

15. Mount Wanhua Peony Garden, Shaanxi

This garden is opposite Huayuantou Village, 15km west of Yan'an of Shaanxi Province. There were visitors to this garden as early as the Song Dynasty, according to historical records. Before 1949, people from scores of kilometres around would go to the temple fair and enjoy the flowers on April 8th of the lunar calendar. The total area on the mountain is 126 hm^2, where 30 000 plants of 12 local peony varieties are divided into five groups: yellow, purple, red, pink and white. The flowers centred around Temple Cuifujun were frequently visited and

Peony Garden in Shanghai Botanical Garden

appreciated because of their unique arrangement among green pines and verdant cypresses. In 1939 and again in 1940 the leaders of the Chinese Communist party Mao Zedong and Zhou Enlai went there to view the tree peonies.

16. Peony Garden in Lake Tian-jing Park of Tongling, Anhui

It was built initially in 1986 in Tongling, Anhui Province. On 1hm^2 of cultivated land nearly 10 000 peonies of 140 varieties, including over ten local varieties, are planted. A natural arrangement based on the terrain has been adopted, creating an abundance of interesting scenery. It is an important centre in Anhui for the appreciation of peonies, and the peony is the city flower of Tongling.

17. Peony Garden in Shanghai Botanical Gardens

Situated in Longhua in Shanghai, this garden covers an area of 4.3hm^2. There are over 7 000 peonies of almost 100 varieties, mainly from Heze in Shandong, Lanzhou in Gansu and Tongling in Anhui. Elegant and unique pavilions, corridors and halls provide visitors with a beautiful environment of appreciation of the plants. Foreign and domestic visitors gather here during flowering, as it is the best place in Shanghai to view peonies. Since 1993, the annual Shanghai Peony Fair has been a welcome attraction to travellers, and it has also brought good social benefits. In addition, peony seedlings have been exported to Holland, USA, Italy and France.

18. Dry Twig Peony Garden, Jiangsu

Located south of Yancheng in Biancang County of Jiangsu Province, this garden has a history of 700 years. The 0.7hm^2 is divided into two parts, the East Garden and the West Garden. In the West Garden there are more than ten red or white peonies that were planted in the Dynasties of Song and Yuan. One tree is as high as 1.9m with a crown diameter of 3.5m, and a trunk diameter, thickest at the base, of 9cm. Nearly 200 flowers grow on it. The branches of these remarkable plants have the appearance of old dry wood. This has given rise to the name Dry Twig Peony. A couplet specially written by General Zhang Aiping of the People's Republic of China is shown on the hall pillars: "Sea water three thousand *Zhang*, Peony flowers seven hundred years." In the East Garden there is a collection of over 500 peony plants in more than 70 excellent varieties from Heze and Luoyang.

19. Double-Pagoda Templ Garden, Shanxi

The garden is situated in the Double-Pagoda Temple in the suburbs of Taiyuan in Shanxi Province. More than ten peonies of the cultivar 'Zi Xia Xian' from the Ming Dynasty have grown there for over 300 years Their branches and twigs look hoary and old but grow luxuriantly. The 2m high trees flower profusely. Approximately 100 varieties have been introduced from Luoyang and Heze. At flowering time, the lush, splendid peonies and the peacefully secluded temple uniquely complement each other in radiance and elegant beauty. It is an important place in Taiyuan for appreciating peonies.

There are many other famous peony gardens in the People's Republic of China. There is not enough space to introduce them one by one in detail. They include: Peony Garden, Beijing Botanical Gardens of Chinese Academy of Sciences; Peony Garden in Huagang Guanyu (Watching Fishes at Flowery Harbour) in Hangzhou; Nanji Peony Garden in Ningguo County, Anhui Province; Hongyuan Peony Garden and Dagongbei Mosque Peony Garden in Zhaotong and Mount Lion Peony Garden in Wuding County of Yun'nan Province; Peony Garden in Eternity Temple, Mount Emei in Sichuan Province; Mount Muoshan Peony Garden in Wuhan in Hubei Province; Peony Garden in Xingqing Park, Xi'an of Shaanxi Province; the Sun Yatsen Mausoleum Peony Garden in Nanjing, Jiangsu Province; Flower Nursery Peony Garden in Bijie, Guizhou; and Peony Garden in Mount Alishan in Taiwan Province.

THE CRITERIA FOR RECORDING MORPHOLOGY OF CULTIVARS

The exterior form of Chinese tree peonies differs from variety to variety, but they share many similarities. They are all shrubs with pinnate leaves; the structure of their flowers is the same; so are their mode and regulation in the aromorphosis of flower forms. The Central Plains group, contains the most varieties and has the widest distribution and richest range of forms. It is therefore selected as a basis and reference group for the detailed description of each morphological recording criterion in the investigation, classification and recording of varietal groups.

Plant Forms

Tree peony plants naturally form a thicket of stems from ground level. Because of the differences in angular divergence of branches, length between branch nodes, and yearly growth of new shoots, the forms and heights of plants vary. They are categorised as follows.

1. Plant Types
There are 3 types.
(1) Erect Type The branches spread straight upward forming small angles generally less than 30°. The distance between nodes on the annual shoots is comparatively long. Because of the rapid and vigorous growth, the plant thickets are tall and upright(fig.12).
(2) Expanding Type The branches expand out diagonally forming angles between 50° and 60°, so that the plant width is significantly greater than its height. Because of the slower rate of growth, the length between branch nodes is shorter and the overall yearly growth is smaller. Its stout thicket is appropriate for pot planted, formal cultivation. Notable among Central Plains cultivars, the height of 'Zi Yi Guan' is less than 20cm after five years' growth and the yearly growth of annual shoots of 'Luo Han Hong' does not exceed 2 or 3cm. In these examples the growth is very slow and the thicket is extremely short(fig.13).
(3) Semi-Expanding Type The angular divergence of branches, the distance between nodes and yearly branch productivity in this type are all intermediate between the previous two types. Since the branches grow at a slant, a semi-expanding plant thicket is produced (fig.14).

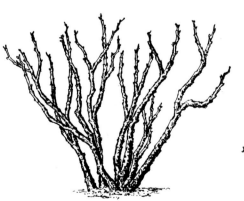

Fig. 12 Erect type

Fig. 13 Expanding type

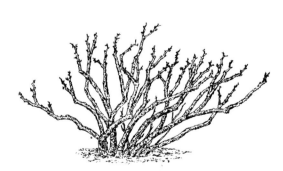

Fig. 14 Semi-expanding type

2. Plant Heights

There are three types, based on the average height of five- to eight-year-old plants.

Tall Type: plant height greater than 80cm.

Medium Type: plant height between 40 and 80cm.

Dwarf Type: plant height less than 40cm.

The criteria of both plant form and height should be described with chronology in cultivar investigation and recording (see details in the introduction to varieties).

Branch Properties

The branch properties are divided into three types for recording and description(fig.15).

1. Old Branches (perennial branches)
In general, the circumference, hardness of texture and smoothness of bark should be recorded and described.

2. Annual Shoots (branches developed in one year)
The growth quantity in the year and comparative length between nodes should be recorded and described.

3. Descendants (basal shoots)
The word descendant is used here to denote a twig sprouting from ground level or below at the base of the plant. The development of descendants varies greatly between varieties. Some produce descendants readily and luxuriantly, while other strains produce very few. Several dozen can grow from high-yielding plants while low-yielding plants have only two or three or even none. Four levels are defined for recording: Abundant, Many, Few and Scarce.

Bud and Shoot Shape and Colour

The shapes and colours of shoot buds and new shoots differ dramatically and are full of changes. They are always the main criteria for distinguishing varieties in winter and early spring. They are also beautiful, attractive and interesting features for appreciation and enjoyment in peony gardens at that time of year. The following are important characters for description:

1. Shape of Shoot Buds
They are classified into the following types: Spherical, Ellipsoidal, Narrow prolate and Prolate(fig.16).

The states of and changes to the tips of the shoot scales, eg whether they are open or curling, are obvious and

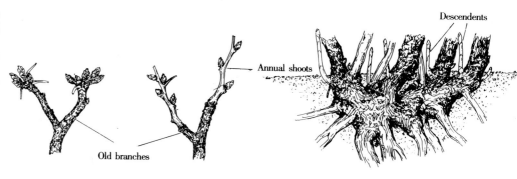

Fig.15 Branch properties

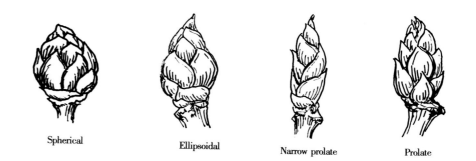

Fig.16 Shape of shoot buds

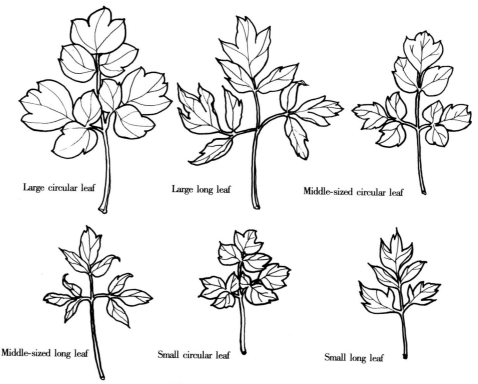

Fig.17 Types of compound leaves

prominent characters for the identification of some varieties.

2. Colours of New Shoots
New shoots after their early spring sprouting show gorgeous colours and plenty of variety. The shoot colours of many cultivars are apparently related to their flower colours, in that a dark-coloured shoot develops a flower of dark colour, or a bud with light colour results in a flower of light colour. On the other hand, a few varieties do not illustrate this relationship. Some even demonstrate a great difference between the shoot colour and the flower colour.

Foliage and Leaf-shapes

Distinctions exist among cultivars in the following aspects: runcinate frequency, sizes of pinnately compound leaves, texture, attachment angles, and shapes and colours of leaflets. These are essential morphological features for cultivar description and therefore should be carefully recorded.

1. Types of Compound Leaves
Most Chinese peony strains possess pinnule, while some develop only pinna. The leaf form should be noted down first. Compound leaves are grouped into one of six types of according to their total length (from the tip of the terminal leaflet to the base of the petiole), total width (the distance between the tips of the outer leaflets on opposite sides of the petiole) and the shapes of leaflets(fig.17):

(1) **Large Circular Leaf** This is a compound leaf large in size with total length greater than 40cm and width greater than 25cm. The leaflets are wide, roundish and fleshy and egg-shaped or wider. A lateral leaflet is flat with fewer incisions on its edge. 'Shou An Hong', 'Mo Kui' and 'Wang Hong' are examples of this type.

(2) **Large Long Leaf** The leaf size is the same as that of a large circular leaf. Its leaflet is long oval or egg-shaped, and thin in texture. There are fewer but more tapering incisions at the lateral leaflet edge. Examples are the leaves of 'Bing Ling Zhao Hong Shi' and 'Yin Fen Jin Lin'.

(3) **Middle-sized Circular Leaf** The size of the compound leaf is intermediate, with total length between 30 and 40cm and width of approximately 20 to 25cm. The leaflet is also an intermediate size. Its other shapes are all similar to those of a large circular leaf. 'Yu Guo Tian Qing' and 'Fen Mian Tao Hua' are in this group.

(4) **Middle-sized Long Leaf** The size of this compound leaf is the same as that of the middle-sized circular leaf,

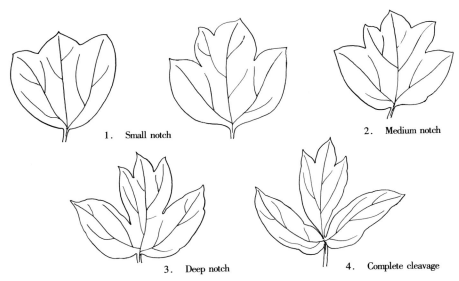

1. Small notch 2. Medium notch 3. Deep notch 4. Complete cleavage

Fig. 18 Apical notches of top leaflets

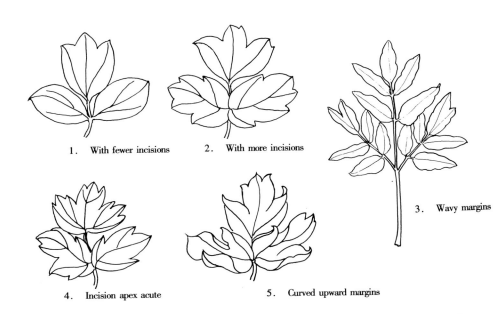

1. With fewer incisions 2. With more incisions 3. Wavy margins 4. Incision apex acute 5. Curved upward margins

Fig. 19 Shapes of lateral leaflets and their edges

1. Roundish apex 2. Sudden tapering apex 3. Sharp tapering apex 4. Gradual tapering apex

Fig. 20 Shapes of leaflet apex

The Criteria for Recording Morphology of Cultivars 43

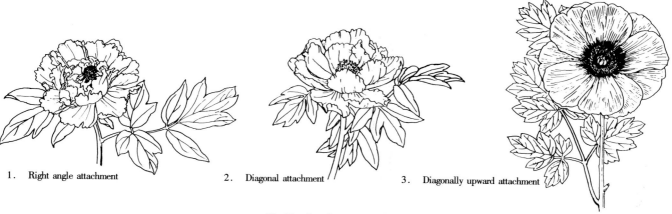

Fig. 21 Attachment of leaves

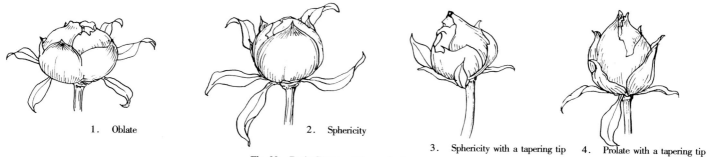

Fig. 22 Basic Shapes of flower buds

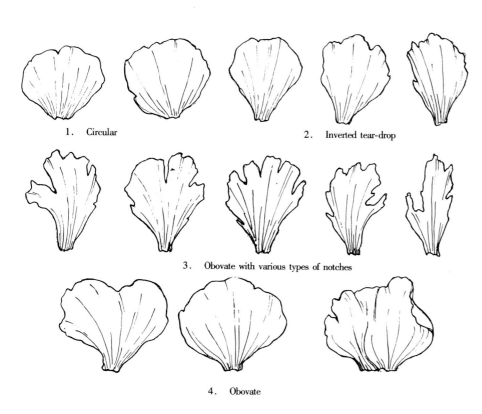

Fig. 23 Shapes of outer petals

while the shape of its leaflet is similar to that of the large long leaf, ie long oval, but with more and sharper incisions. 'Jia Ge Jin Zi' and 'Lu Fen' are of this type.

(5) **Small Circular Leaf** The compound leaf of this type is small in size, ie its total length is about 20 - 30cm and the width does not exceed 20cm. The leaflet of this type is also small in size. Their shapes are the same as described for other circular leaves. 'Lan Tian Yu' and 'Chi Long Huan Cai' are in this category.

(6) **Small Long Leaf** The size of its compound leaf and leaflet is similar to that of the small circular leaf. Its leaflet shows the same shape as that of other long leaves. Examples are 'Ying Luo Bao Zhu' and 'Tao Hong Xian Mei'.

2. Apical Notches of Top Leaflets (fig.18)
(1) Small notch
(2) Medium notch
(3) Deep notch
(4) Complete cleavage

3. Shapes of Lateral Leaflets and Their Edges (fig.19)
(1) With fewer incisions
(2) With more incisions
(3) Wavy margins
(4) Incision apex acute
(5) Curved upward margins

44 Chinese Tree Peony

4. Shapes of Leaflet Apex(fig.20)
(1) Roundish apex
(2) Sudden tapering apex
(3) Sharp tapering apex
(4) Gradual tapering apex

5. Attachment of Leaves
The compound leaves are attached to the stem at various angles, and petiole lengths vary. The attachment angles can be usefully classified into three types(fig.21):
(1) Right angle attachment (approximately 90°)
(2) Diagonal attachment (30° to 50°)
(3) Diagonally upward attachment (less than 30°).

Basic Shapes of Flower Buds (fig.22)

1. Oblate
2. Sphericity
3. Spherical with a tapering tip
4. Prolate with a tapering tip

Shapes of Outer Petals (original, exterior petals)(fig.23)

1. Circular
2. Inverted tear-drop
3. Obovate with various types of notches
4. Obovate

Shapes of Calyces and their Heteromorphosis(fig.24)

1. Original (normal) shapes
2. Heteromorphous shapes after calycanthemy (exterior coloured sepals)

Shapes of Stamens and Varieties after Calycanthemy (fig.25)

Shapes of Pistils and Varieties after Calycanthemy (fig.26)

Shapes of Fruits and Seed (fig.27)

Flower Forms

1. Typical Flower Forms(fig.28)
(1) Single
(2) Lotus
(3) Chrysanthemum
(4) Rose
(5) Anemone
(6) Golden Circle
(7) Crown
(8) Globular

2. Criteria for Recording Flower Forms
The eight forms listed above are ranked in increasing order of flower

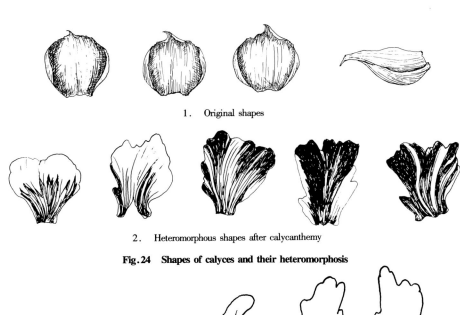

1. Original shapes

2. Heteromorphous shapes after calycanthemy

Fig.24 Shapes of calyces and their heteromorphosis

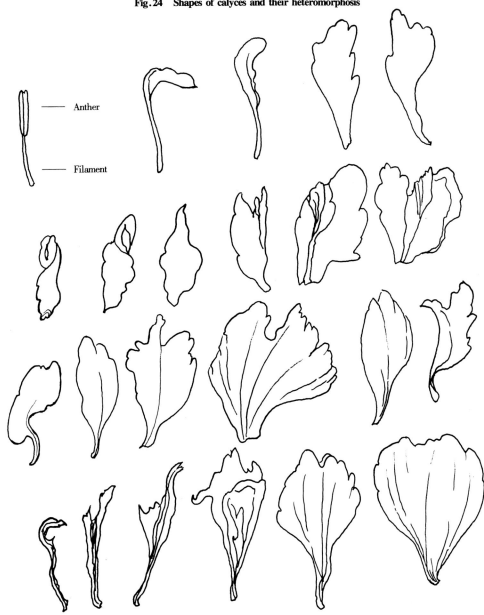

Fig.25 Shapes of stamens and varieties after calycanthemy

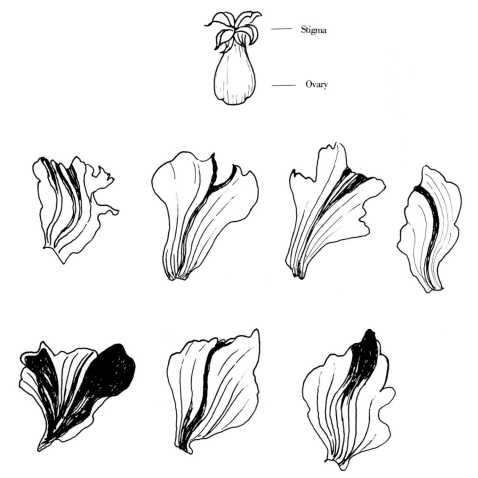

Fig. 26 Shapes of pistils and varieties after calycanthemy

Fig. 27 Fruits and seed

development. The flower form of a cultivar should be determined by its flower form of the highest rank. At least 60% of the flowers must emerge in a stable way and be of a consistent form. Only then can the variety be classified as having flowers of that particular form. Most varieties can be classified in this way. However, some varieties of the Central Plains Group are stably heteromorphous: two or even three flower forms can be found on one plant. For example, the cultivar 'Zhao Fen' chronologically develops flowers of three forms: single, lotus and crown. Heteromorphosis also frequently occurs in plants of the varieties 'Gong Yang Zhuang', 'Qing Xiang Bai' and 'Guan Yin Mian'. Such varieties should be classified, at least temporarily, as 'multi-flowered' forms. Their flowers of superior rank should be regarded as the major form while other forms should also be included in the description. Further research is necessary into heteromorphous plants. As a further complication, the flower forms of some varieties are in fact an intermediate or transitional flower form between two of the above forms, for example intermediate between the hundred proliferate and crown proliferate categories. The flowers of 'Tao Hong Zeng Yan', 'Yao Chi Chun' and 'Hai Tang Zheng Run' show examples of intermediate forms between rose and crown forms so that it is not possible to assign them to one particular category. While recording the morphology of such varieties therefore, a possible solution is to list them as intermediate forms.

3. Criterion for Recording Flower Size

Flower size is denoted by flower diameter × flower depth, in centimetres.

Flower Colours

The colours of Chinese peonies are rich and varied. There are primarily seven mono-chromatic categories denoted by white, yellow, pink, red, purple, blue, and green. There are also the type with two colours, one of which bleeds into the other, and the multi-colour type which is versicolorous. Two points should be emphasised in recording:

1. Standard colours

The standard colour of each cultivar should be determined when the flower is initially in full bloom. For the multi-colour varieties, their colours in the three periods of initial bloom, full bloom and final bloom respectively should be recorded, so that the complete versicolorous process in a particular variety can be described.

2. Description of colours

The colours are described by the conventional terminology of colours in Chinese chromatology. The Colour Chart of the RHS (The Royal Horticultural Society, London, England) should be referred to for consistency of recording and the colour code added in brackets. (For details see the Introduction to Cultivars.)

Florescence

The three flowering periods, early, mid-season and late, are defined below, based on the flower chronology in the locations of Luoyang, Henan Province and Heze, Shandong Province.

Early
From early April to early mid-April.
Mid-season
From mid-April to late April.
Late
From the end of April to early May.
Florescence of a single flower
The complete flowering period of a single flower from initial opening to fading.
Florescence of a colony
The period between 5% of the flowers initially blooming and 95% of the flowers fading.
Florescence of initial bloom
The period when 5% of the flowers are in bloom.
Florescence of full bloom
The period when over 50% of the flowers are in bloom.
Florescence of final bloom
The period when about 80% of the flowers have faded.

See the table of cultivar morphology record (Table 7).

46 CHINESE TREE PEONY

1. Single
2. Lotus
3. Chrysanthemum
4. Rose
5. Anemone
6. Golden Circle
7. Crown
8. Globular

Fig. 28 Typical flower forms

Table 7 Table of Cultivar Morphology Record

Number		Date		Place	
Name of Cultivar		Alias		Parental Combination	
Flower Type		Flower Colour		Flower Size	
State of Flower Stalk		Flowering State		Flowering Time	Early
					Full
					Fading
Shape and Texture of Petal	Outer Petal				
	Inner Petal				
	Stamen				
	Pistil				
Plant Type		Flower Bud	Shape	Characters of Branch	Old Branches
			Colour		Annual Shoots
Foliage and Leaf – shapes		Top Leaf Shape		Lateral Leaf Shape	
Leaf Colour		Leaf Stalk		Petiolule	
Breeding Ability		Branch Ability		Basal Shoots	

Introduction And Development Of Chinese Peonies In Other Countries

Chinese Peonies in Japan

The peony has long been an important plant in Japan in terms of history, extent of cultivation and significance in people's lives. Japan has also played a role alongside China in introducing peonies to the world. It is said that in the eighth century peonies and other plants were carried to Japan by the respected monk Fa Hai. Peonies were first planted in temples. They soon became popular and were planted by local people. There was widespread interest in their medical use, and peony cultivation became a great industry in Yomata and Yamasiro. With exchanges between Chinese and Japanese culture and the export of peony varieties, Chinese peony culture gradually filtered into Japan. The peony became a royal flower, next to chrysanthemum in position, together with sakura and lotus. The initial period of Japanese development of peonies was from 1615 to 1867 when improved varieties of peonies were produced. During this period of cultivation, Japanese peonies developed their own particular features: single or semi-double forms with bright colours and beautiful shapes, and different from the Chinese flowers. Peony cultivation reached its climax in Japan in 1818 - 1830, and many gardens were established in *Tokyo*. After 1868 it reached the industrial horticulture stage and peony nurseries were established.

At the beginning of the twentieth century, peonies were exported from Japan to Europe and America in large numbers. Although this industry stopped in the 1939-1945 War, it has completely recovered now. Over 10% of the two million plants produced every year are sent to the United States.

Chinese Peonies in the West

1. Europe

When visiting Beijing in 1665, people from the East India Company recorded the peonies that they personally saw. A century later, Joseph Banks from Kew Gardens in Britain read these papers and saw Chinese paintings, so in 1787 he asked Alexander Duncan of the same company to collect plants in Guangzhou and send them to Kew. After two years the blooming of a pink double was recorded as *P. moutan* 'Banksii' or 'Powder Ball'. But this, the earliest Chinese peony growing in Europe, was damaged in 1842 during garden reconstruction. Fortunately, in the garden of the Duncan family in Arbroath, a white double with rosy colouring at the petal base was still growing. It was transplanted into the Royal Botanic Garden, Edinburgh, and was surely a direct descendant of the first group of peonies sent to Kew. More peonies were sent from Guangzhou to Kew in 1794, but only a few survived to reproduce and be spread to other places. On the basis of one of these cultivated plants, Andrews described *P. suffruticosa* Andr. as a new species, the first tree species of the genus *Paeonia*.

Another collection of peonies was shipped to Britain in 1802. A pinkish-white flower with reddish purple blotches attracted wide interest. Later seed collection and seedling cultivation suggest that this was the initial introduction of *P. rockii* into Europe. Peonies came to Britain, France and other European countries from 1787 by various routes. The British botanist Robert Fortune collected hundreds of peony plants in China, but when they flowered there seemed to be only five varieties. Another expedition during 1843-1845 to Shanghai for more plants, arranged by the British Royal Horticulture Association, resulted in complete failure. However, on his third trip from 1848 to 1851, Fortune obtained in Shanghai 30 varieties of Mudan said to be the best at the time, and some Shaoyao for grafting. This introduction was the start of a major expansion, and by 1860 most British nurseries sold these varieties.

The second half of the 19th century was the heyday for Chinese peonies in Europe. In the 1860s the nurseries of Heage and Schmidt in Germany, Krelage in Holland, and Verdier in France sold scores or hundreds of varieties. In the next ten years the Belgian Van Houtte and the German Spaeth listed some hundreds of varieties but without details. Paillet Nursery near Paris listed 337 in the 1890s. Most of the better varieties were indicated to be of Italian origin, but there was no reference to show whether they were grown from seed or imported directly from China or Japan. This widespread cultivation demonstrates that Chinese peonies had adapted well to the European environment.

2. USA

It is not known when tree peonies were first brought to the United States. But the earliest appearance of the plant may have been around 1820 when it was introduced from England, rather than directly from China. In 1826, W. Lathe imported cv. *paveracea* from Britain. Two years later in some nurseries a small number of peonies were sold at $5 each, very expensive at the time. Perkins in Massachusetts directly introduced the Chinese cv. Rawsei in 1836 and at the same time

Rawsei in 1836 and at the same time another person obtained some Mudan seedlings from France. Then Prince introduced 20 varieties from Europe to Flushing at a high cost. In contrast to their popularity in Europe at the beginning of the 20th century, only a few American parks exhibited peonies. However, the establishment of the American Peony Society in 1904 and the publication in 1919 of *The Book of the Peony* by Alice Harding greatly boosted the spread of peonies in America. The society is now the international registration organization for peonies, appointed by the International Cultivated Plants Registration Office. The *APS Bulletin* has the aims of transferring knowledge, facilitating exchanges of varieties and encouraging interest. To date nearly 300 issues have been published. It also serves as a record of the development of peonies in the USA, and has a degree of influence elsewhere.

In the period before the 1914-1918 War, large quantities of Chinese peonies were imported into the USA from British, French and Dutch wholesalers. Japanese peonies were introduced to the USA at the same time. Therefore at present the varieties in the country are a mixture of cultivars bred in the USA, cultivars which came from Europe (some are probably Chinese varieties that were renamed) and Japanese cultivars. There are few Chinese varieties explicitly acknowledged as such, but recently plants have been imported directly from China by Cricket Hill Garden, for example, who have established a newsletter entitled 'Peony Heaven' to provide American people with information about Chinese peonies.

A. P. Saunders, the well known American thremmatologist, introduced plants like *P. delavayi* and *P. lutea* to New York and, by hybridising them with Japanese cultivars, produced a series of famous cultivars, for example 'Black Pirate' and 'Age of Gold'. More recently other hybridisers including W. Gratwick and N. Daphnis have continued this work. Thus, from Saunders' pioneering developments, American peony breeding has become very active, and 54 cultivars were officially registered by seven hybridisers in the decade 1976-1986.

3. Chinese Wild Species Peonies in the West

(1) *P. delavayi* and *P. lutea* In the 1880s the French missionary A. Delavay found *P. delavayi* and *P. lutea* in Yunnan, and by 1892 they were in bloom in France. L. Henry and M. Lemoine later used them to produce the yellow Lemoine hybrids.

(2) *P. lutea* var. *ludlowii* Stern and Taylor This subspecies was found by Ludlow and Sherriff in Zangbu Valley in south-east Tibet in 1936. The flower was initially regarded as *P. lutea* until 1953 when it was recognised as a distinct taxon. It is considered good breeding material and reliable for seed production. Some new hybrids were successfully produced in Britain after hybridisation with *P. delavayi* and other varieties.

(3) *P. potaninii* Komar It was rediscovered and named by the botanist Potanin in Yunnan and Sichuan in 1921. The plant was in fact introduced to Britain by Wilson and flowered there in 1911. A subspecies, *P. potanini* var. *trollioides* (Stapf and F. C. Stern) was introduced to Britain in 1914 by Forrest. It has not produced the useful hybrids that might be expected.

(4) *P. spontanea* T. Hong and W. Z. Zhao This peony, first known as a cultivated plant, is an important ancestral species for cultivation. Although its wild distribution was popularly recorded in ancient literature, its form, properties and growth habits were not described in detail. William Purdom found the wild plant near Yan'an, Shaanxi Province, in 1910 and 1911, and named it *P. spontanea*. He gave a sample to the Arnold Arboretum, and seeds to Veitch Nursery in Britain and Professor Sargent at Harvard University. Unfortunately few plants grew well, and because of its relatively small and unspectacular flowers the species has not become well established in Europe and America.

(5) *P. rockii* T. Hong and J. J. Li Reginald Farrer found examples of *P. rockii* in south-west Gansu in 1914, but he did not collect samples. During 1925-1926 Joseph Rock collected seeds from a cultivated plant in a Lama temple garden and sent them to the Arnold Arboretum, so that what was introduced to America and to European countries was a group of cultivar seedlings. In 1938 plants were blooming successfully in North America, Canada, Britain and Switzerland. The British peony scholar F. C. Stern received a seedling in 1936. After two years it flowered and by 1959 it had reached 2.5m in height and 3.7m in breadth. Plants developed from *P. rockii* and its cultivars are collectively called 'Rock's Variety' in the West. They are a much sought-after group of plants, greatly admired, and have been used for breeding. Examples of *P. rockii* and of hybrids derived from it are now growing in some nurseries and gardens in North America and Europe. Since it cannot be reproduced rapidly however, *P. rockii* and its cultivars are not as common as Saunders varieties and Japanese varieties.

(6) *P. ostii* T. Hong and J. X. Zhang This is the ancestral species of *Fengdan* peonies which are widely cultivated for medicinal purposes. It has been recently introduced from Zhengzhou to Italy by G. L. Osti. The closely related varieties 'Fengdanbai' and 'Fengdanfen' have also been sent to Italy and Britain and doubtless will spread to other European and American countries.

CULTIVAR STATEMENTS

Single Form

1–3 whorls of normal petals that are wide, large and flat with a wide-ovate, ovoid or obovate shape. Stamens normal and pistils normal and fertile.

cv. Bai Yu Lan

Single form. Flowers 16cm × 6cm, white (155-D); petals 2-whorled, broad and large, entire and regular, light pink at the base; stamens normal, filaments pale purple; pistils normal; floral discs pale purple. Stalks straight, flowers upright. Flowering early.

Plant medium height, erect. Branches stout. Leaves orbicular, medium-sized, crowded; leaflets ovate, acuminate at the apex, slightly curved downwards on the margins. Growth vigorous, flowers many. Bred by Wangcheng Park, Luoyang in 1969.

Bai Yu Lan

Yu Lan Piao Xiang

cv. Yu Lan Piao Xiang

Single form. Flower buds orbicular; flowers 21cm × 6cm, white (155-D), lustrous; petals large, entire and regular, stiff, suffused with pale purple at the base; stamens normal or slightly petaloid; pistils normal, stigmas yellow; floral discs pinkish purple. Stalks straight, flowers upright. Flowering midseason.

Plant tall, erect. Branches stout. Leaves orbicular, large-sized, sparse, thick; leaflets ovate, apex obtuse, the margins curved downwards, suffused with purplish red. Growth vigorous, flowers many. Bred by Luoyang Peony Garden, Luoyang in 1995.

54 Chinese Tree Peony

Chi Tang Xiao Yue

cv. Chi Tang Xiao Yue

Single form, sometimes anemone form occurs. Flowers 17cm × 9cm, creamy yellow (158-D), lustrous; petals 2-3-whorled, with dark purple streaks at the base; stamens partially developed, petaloid, slender and long, filaments normal; anthers developed, petaloid, small and curved; pistils normal; floral discs white. Stalks fairly straight, flowers lateral. Flowering midseason.

Plant medium height, partially spreading. Branches relatively stout. Leaves orbicular, medium-sized, sparse; leaflets long elliptic with numerous lobes on the margins, apex acute. Growth medium, number of flowers normal. Classic variety.

cv. Chi Yang

Single form. Flowers 15cm × 4cm, red (57-D); petals 3-whorled, broadly ovate, dark purple basal blotches; stamens normal; pistils normal; floral discs purplish red. Stalks slender, flowers upright. Flowering midseason.

Plant dwarf, erect. Branchlets fairly long. Leaves orbicular, small-sized; leaflets elliptic, apex acute. Growth medium, flowers many. Bred by Wangcheng Park, Luoyang in 1995.

Chi Yang

Single Form

Pan Zhong Qu Guo

Ying Su Hong

cv. Pan Zhong Qu Guo

Single form. Flowers 12cm × 4cm, faintly purple (64-C); petals 1-2-whorled, purplish red basal blotches; stamens normal; pistils normal, floral discs purplish red. Stalks fairly long, flowers upright. Flowering early.

Plant medium height, partially spreading. Branches slender, stiff. Leaves long, small-sized, sparse; leaflets long ovate, apex acute; upper surface yellowish green, suffused with purplish red. Growth vigorous, flowers many. Classic variety.

cv. Ying Su Hong

Single form. Flower buds conical; flowers 13cm × 4cm, pale purplish red (67-C); petals 2-3-whorled, obovate, stiff, lustrous, purple basal blotches; stamens normal; pistils normal, stigmas purplish red; floral discs purplish red. Stalks relatively short, flowers slightly hidden among the leaves. Flowering midseason.

Plant dwarf, partially spreading. Branches fairly slender. Leaves orbicular, medium-sized; leaflets long elliptic or ovate, apex acute; upper surface smooth, green. Growth medium, flowers many.

Hei Hai Jin Long

cv. Hei Hai Jin Long
Single form. Flowers 16cm × 3cm, purplish red (187-D), soft; petals fairly stiff, suffused with dark purple at the base; stamens normal; pistils normal; floral discs deep purple. Stalks slightly long, stiff, straight. Flowering early.
Plant dwarf, erect. Branches relatively stout, stiff. Leaves orbicular, medium-sized, stiff; leaflets broadly ovate, apex obtuse, the margins curved upwards. Growth medium, flowers many. Bred by Zhaolou Peony Garden, Heze in 1964.

cv. Fei Yan Zi
Single form. Flowers 16cm × 3cm, purplish red (61-C); petals 3-whorled, broadly ovate, dark purple basal blotches; stamens normal; pistils normal. Stalks slender, straight. Flowering midseason.
Plant dwarf, partially spreading. Leaves orbicular, small-sized; leaflets elliptic, apex acuminate, terminal leaflets deeply cut. Growth fairly vigorous, flowers many. Bred by Wangcheng Park, Luoyang in 1995.

cv. Zi Die
Single form. Flowers 16cm × 5cm, purplish red (61-C); petals 2-3-whorled, broad and large, broadly ovate, suffused with deep colour, stamens normal; pistils normal. Stalks slender, long. Flowering midseason.
Plant medium height, partially spreading. Branchlets long. Leaves orbicular, medium-sized; leaflets orbicular, slightly curved upwards on the margins. Growth vigorous, flowers many. Bred by Wangcheng Park, Luoyang in 1995.

cv. Zao Yan Hong
Single form. Flowers 18cm × 6cm, purplish red (61-C); petals 3-whorled, broad and large, entire and regular, relatively stiff; stamens normal; pistils normal, stigmas purplish red; floral discs purplish red. Stalks relatively slender, flowers upright. Flowering early.
Plant medium height, partially spreading. Branches slender, stiff. Leaves orbicular, medium-sized; leaflets ovate, apex acute, the margins curved upwards. Growth vigorous, flowers many. Bred by Luoyang Peony Garden, Luoyang in 1995.

Fei Yan Zi

Zi Die

Zao Yan Hong

cv. Cai Die

Single form. Flowers 18cm × 3cm, purplish red (53-B) with pink; petals 2-whorled, broad and large, stiff; slightly wrinkled, with white radical streaks, bright purplish red at the base; stamens normal; pistils normal; floral discs creamy white. Stalks slender, stiff, fairly long, flowers upright. Flowering early.

Plant medium height, erect. Branches slender, stiff. Leaves orbicular, medium-sized, stiff, sparse; leaflets ovate or long ovate, acuminate at the apex, the margins slightly curved downwards. Growth medium, flowers many. Bred by Zhaolou Peony Garden, Heze in 1995.

Cai Die

cv. Zi Jin He

Single form or lotus form. Flowers 15cm, purplish red (53-B); petals 2-3-whorled, soft, flat, colour pure, the outer and the inner petals of same colour, lustrous; stamens normal; pistils normal; floral discs pale purple. Stalks stout, stiff, slightly short, flowers upright and slightly lateral. Flowering early.

Plant medium height, partially spreading. Branches stout, purple blotches at the leaf axils. Leaves orbicular, medium-sized; leaflets ovate or broadly ovate, suffused with purple on the margins. Growth vigorous, resistant to root diseases, flowering period early. Bred by Zhaolou Peony Garden, Heze in 1990.

Zi Jin He

LOTUS FORM

Large and neat petals in 4 – 5 slightly overlapping whorls forming the shape of a lotus flower. Stamens normal and pistils normal.

cv. Hai Tang Hong

Lotus form. Flowers 18cm × 4cm, deep pinkish red (52-D); petals 4-5-whorled, broad and large, entire and regular, fairly stiff, the outer ones and the inner ones of the same colour, fresh, soft and charming; stamens normal; pistils normal; floral discs deep purplish red. Stalks fairly long, straight, flowers upright. Flowering early.

Plant medium height, partially spreading. Branches stout. Leaves long, medium-sized, horizontal; leaflets long ovate, obliquely ascending, acuminate at the apex, the margins slightly curved upwards, slightly suffused with purple. Growth vigorous, flowers many, flowering period long. Bred by Zhaolou Peony Garden, Heze in 1991.

Hai Tang Hong

Yu Ban Bai

Huang Hua Kui

Yu Yi Huang

cv. Yu Ban Bai

Lotus form. Flowers 17cm × 4cm, white (155-D); petals entire and regular, stiff, slightly suffused with pink at the base; sometimes stamens petaloid, pistils normal; floral discs pinkish white. Stalks slightly long, stiff, flowers upright. Flowering early.

Plant dwarf, erect. Branches slender, stiff. Leaves long, medium-sized, sparse; leaflets long ovate, acuminate at the apex, the margins slightly curved upwards. Growing slowly, but flowers many. Classic variety.

cv. Huang Hua Kui

Lotus form, sometimes stamens petaloid. Flowers 14cm × 5cm, yellowish (158-D), suffused with pale purple; petals stiff, purple basal blotches; stamens normal or petaloid; pistils normal, very fertile; floral discs white. Stalks long, flowers upright. Flowering early.

Plant tall, erect. Branches slender, stiff. Leaves orbicular, medium-sized, stiff, sparse; leaflets nearly orbicular, apex obtuse, the margins curved upwards. Growth vigorous, flowers many. Classic variety.

cv. Yu Yi Huang

Lotus form or chrysanthemum form. Flowers 15cm × 5cm, yellowish (19-D); petals broad and large, wavy, stiff, suffused with pale purple at the base; stamens occasionally petaloid; pistils normal; floral discs yellowish, very fertile. Stalks slightly long, flowers upright. Flowering midseason.

Plant medium height, erect. Branches relatively slender, stiff. Leaves orbicular, medium-sized, thick, sparse; leaflets ovate or broadly ovate, apex obtuse. Growth weak; flowers relatively few. Classic variety.

cv. Zi Hong Ling

Lotus form. Flowers 18cm × 4cm, purplish red (53-B), lustrous; petals fairly thick, stiff, with roundish teeth and rugose at the apex, dark purple basal blotches; stamens normal; pistils normal, very fertile; floral discs deep purplish red. Stalks short, stiff, flowers upright. Flowering midseason.

Plant medium height, partially spreading. Branches stout. Leaves orbicular, large-sized, thick; leaflets ovate, apex obtuse, suffused with purple on the margins. Growth vigorous, flowers many. Bred by Zhaolou Peony Garden, Heze in 1964.

cv. Hong Yan He

Lotus form, sometimes chrysanthemum form occurs. Flowers 15cm × 4cm, light red (52-A); petals 4-5-whorled, long elliptic, dark purple basal blotches; stamens normal; pistils normal; floral discs purplish red. Stalks stout, straight, flowers upright. Flowering midseason.

Plant medium height, partially spreading. Branchlets fairly long. Leaves long, medium-sized; leaflets long elliptic or lanceolate, acuminate at the apex. Growth medium, flowers many. Bred by Wangcheng Park, Luoyang in 1995.

cv. Jiao Nu

Lotus form. Flowers 17cm × 5cm, at first light red somewhat pink (52-D), lighter colour when fully open; petals 4-whorled, ovate, stamens normal; pistils developed, petaloid, greenish coloured. Stalks stout, straight, red. Flowering midseason.

Plant medium height, erect. Leaves orbicular, medium-sized; leaflets elliptic, acuminate at the apex, the margins curved upwards, veins sunken. Growth medium, flowers many. Bred by Wangcheng Park, Luoyang in 1994.

cv. Da Ban Hong

Lotus form, sometimes chrysanthemum form occurs. Flowers 22cm × 5cm, deep purplish red (61-C); petals 4-6-whorled, entire, large, flat, apex toothed, thick, stiff, dark purple basal blotches; stamens occasionally petaloid; pistils 11, floral discs pale purplish red. Stalks fairly long, stiff, flowers lateral. Flowering midseason.

Plant medium height, partially spreading. Branches stout. Leaves long, medium-sized, crowded; leaflets long ovate or broadly lanceolate, acuminate at the apex, pendulous, the margins wavy and slightly curved upwards. Growth medium, flowers many. Bred by Zhaolou Peony Garden, Heze in 1979.

cv. Da Jin Fen

Lotus form or chrysanthemum form. Flowers 15cm × 6cm, pinkish purple (65-A); petals 4-8-whorled, broad and large, entire and regular, suffused with deep colour at the base; stamens normal, occasionally petaloid; pistils normal, stigmas purplish red; floral discs purplish red. Stalks relatively slender, flowers lateral. Flowering midseason.

Plant medium height, spreading. Branches relatively slender. Leaves long, medium-sized; leaflets lanceolate or ovate, acuminate at the apex. Growth medium, flowers many. Classic variety.

cv. Hong Lian Man Tang

Lotus form, sometimes crown form occurs. Flowers 16cm × 7cm, light red (58-C), lustrous; petals 3-4-whorled, large, with a deep-cut lobe at the apex, stiff, purple at the base; stamens normal or occasionally petaloid; pistils normal; floral discs purplish red. Stalks erect. Flowering midseason.

Plant tall, erect. Branches stout. Leaves long, large-sized, thick; leaflets long elliptic, apex acute, the margins curved upwards. Growth vigorous, flowers many. Bred by Wangcheng Park, Luoyang in 1969.

cv. Gu Ban Tong Chun

Lotus form. Flowers 18cm × 4cm, pinkish white (36-D); petals stiff, arranged regularly, apex shallowly toothed, purple at the base; stamens normal, occasionally petaloid, pistils normal; very fertile. Stalks stiff; flowers upright. Flowering midseason.

Plant medium to dwarf, partially spreading. Branches stout. Leaves long, medium-sized, stiff, fairly sparse; leaflets long ovate, acuminate at the apex. Growth vigorous; flowers many. Classic variety.

cv. Fen E Jiao

Lotus form. Flowers 15cm × 6cm, light pink bluish (65-B); petals gradually shorter from outer to inner whorls, arranged regularly, deep purple basal blotches; stamens normal; pistils normal; floral discs pink, stigmas yellow. Stalks straight, flowers upright. Flowering early midseason.

Plant tall, erect. Branches slender, stiff. Leaves orbicular, medium-sized; leaflets long ovate, thick, the margins slightly curved upwards. Growth medium, flowers normal. Classic variety.

Zi Hong Ling

Hong Yan He

Jiao Nü

Da Ban Hong

Lotus Form 63

Da Jin Fen

Hong Lian Man Tang

cv. Xian Chi Zheng Chun

Lotus form, sometimes chrysanthemum form occurs. Flowers 18cm × 4cm, pinkish white (36-D); petals stiff, layered conspicuously, apex toothed, suffused with pink at the base; stamens normal; pistils normal or occasionally petaloid. Stalks stout, stiff, flowers upright. Flowering midseason..

Plant medium height, partially spreading. Branches stout, stiff. Leaves long, medium-sized, stiff, sparse; leaflets ovate, acuminate at the apex, terminal leaflets raised. Growth vigorous; flowers many. Classic variety.

cv. Chun Lian

Lotus form. Flowers 15cm × 4cm, pinkish red (62-C); petals 4-whorled, fairly thin, wavy and wrinkled, toothed on the margins, suffused with red at the base; stamens normal; pistils normal; floral discs bright purplish red. Stalks fairly short, flowers upright and slightly lateral. Flowering early.

Plant dwarf, partially spreading. Branches slender, stiff. Leaves long, small-sized; leaflets ovate or long ovate, acuminate at the apex, slightly pendulous. Growth fairly vigorous, flowers many. Bred by Zhaolou Peony Garden, Heze in 1990.

cv. Jiu Zui Yang Fei

Lotus form or anemone form, sometimes crown form occurs. Flowers 25cm × 8cm, pinkish purple (73-B), the outer petals 2-whorled, large, soft, purplish red at the base; stamens partially petaloid; pistils regular. Stalks long, pliable, flowers lateral or pendulous. Flowering midseason.

Plant tall, spreading. Branches stout, pliable, curved. Leaves long, large-sized, thick, soft, extremely sparse; leaflets ovate or long ovate, acuminate at the apex, pen-dulous. Growth vigorous; flowers many. Classic variety.

cv. Hong He

Lotus form. Flowers 15cm × 4cm, pale purplish red (57-B); petals 4-whorled, fairly thin, suffused with red at the base, stamens normal; pistils normal, fertile; floral discs deep purplish red. Stalks stiff, flowers upright. Flowering midseason.

Plant medium height, partially spreading. Branches stiff. Leaves orbicular, medium-sized, slightly sparse; leaflets ovate, apex acute, the margins slightly wavy upwards, suffused with purple. Growth medium, flowers many. Bred by Heze Flower Garden in 1980.

Gu Ban Tong Chun

Fen E Jiao

Xian Chi Zheng Chun

Chun Lian

Jiu Zui Yang Fei

Hong Xia

Mo Sa Jin

Zi Xia Xiang Yu

Zi Xia Xian

Hong He

cv. Hong Xia

Lotus form. Flowers 17cm × 3cm, purplish red (63-B); petals 3-whorled, entire and regular, stiff, suffused with purple at the base; stamens normal; pistils normal, very fertile; floral discs pale purplish red. Stalks slender, long, stiff, flowers upright or lateral. Flowering midseason.

Plant tall, erect. Branches slender, stiff. Leaves orbicular, small-sized, stiff, sparse; leaflets ovate, apex acute. Growth vigorous, flowers profuse. Bred by Heze Flower Garden in 1980.

cv. Mo Sa Jin

Lotus form. Flowers 14cm × 4cm, dark purplish red (187-B), bright; petals soft, irregularly toothed at the apex, black basal blotches; stamens normal; pistils normal; floral discs dark purple. Stalks slender, pliable, long, flowers laterally pendulous. Flowering midseason.

Plant medium to dwarf, erect. Branches slender, stiff. Leaves long, medium-sized, soft, sparse; leaflets long ovate or long elliptic, acuminate at the apex, pendulous, suffused with purplish red on the margins. Growth weak, flowers many. Classic variety.

cv. Zi Xia Xiang Yu

Lotus form. Flowers 16cm × 4cm, purple (63-C), pinkish purple at the apex; petals thin, soft, with conspicuous streaks, numerous teeth on the margins; stamens occasionally petaloid; pistils normal, very fertile. Stalks fairly long, straight, flowers upright. Flowering midseason.

Plant dwarf, erect. Branches relatively slender. Leaves long, large-sized; leaflets long ovate, acuminate at the apex, pendulous, curved upwards on the margins. Growth medium, flowers many. Bred by Zhaolou Peony Garden, Heze in 1963.

cv. Zi Xia Xian

Lotus form. Flowers 12cm × 4cm, purplish red (53-B), lustrous; petals stiff, dark purple basal blotches; stamens normal; pistils normal; floral discs deep purplish red. Stalks stiff, fairly long, flowers upright. Flowering early.

Plant dwarf, erect. Branches slender, stiff. Leaves orbicular, small-sized, sparse; leaflets ovate, stiff, apex obtuse, suffused with pale purple on the margins. Growth weak, flowers many. Classic variety.

Fen Zi Han Jin

cv. Fen Zi Han Jin

Lotus form. Flowers 18cm × 10cm, purplish red (63-B), lustrous; petals 5-6-whorled, broad and large, entire and regular, slightly deep colour at the base, stamens normal; pistils normal; floral discs deep purplish red. Stalks fairly pliable, flowers lateral. Flowering midseason.

Plant medium height, partially spreading. Branches stout. Leaves orbicular, large-sized, thick; leaflets ovate, acute at the apex, the margins slightly curved upwards, with conspicuous veins. Growth fairly vigorous; flowers many. Bred by Wangcheng Park, Luoyang in 1969.

Lotus Form 67

Fen Zi Han Jin

Zhu Sha Kui

Mo Zi Rong Jin

cv. Zhu Sha Kui

Lotus form. Flowers 14cm × 6cm, pinkish purple, gradually deepening in colour towards the centre (73-A); petals large, with a few and large teeth, purple basal blotches; stamens normal; pistils normal, floral discs purplish red. Stalks erect, flowers upright. Flowering midseason.

Plant medium height, erect. Branches slender, stiff. Leaves orbicular, medium-sized, sparse; leaflets ovate, apex obtuse. Growth medium, flowers average. Bred by Wangcheng Park, Luoyang in 1969.

cv. Mo Zi Rong Jin

Lotus form. Flowers 13cm × 6cm, dark purple (187-B), lustrous; petals soft, sparse, suffused with dark purple at the base, a few stamens petaloid; pistils small; floral discs dark purplish red; stalks slender, short; flowers hidden among the leaves. Flowering early.

Plant tall, partially spreading. Branches stout, stiff. Leaves long, medium-sized, fairly sparse; leaflets elliptic, obtuse or acute at the apex, the margins slightly curved upwards. Growth vigorous, flowers few. Bred by Zhaolou Peony Garden, Heze in 1963.

cv. Ba Bao Xiang

Lotus form. Flowers 13cm × 3cm, red (55-A); petals thin, soft, dark purple at the base; stamens normal; pistils normal, very fertile; floral discs purplish red. Stalks short, flowering among the leaves. Flowering midseason.

Plant dwarf, partially spreading. Branches slender. Leaves long, small-sized, thick, sparse; leaflets long ovate, apex acuminate, the margins wavy upwards. Growth medium, flowers many. Classic variety.

Ba Bao Xiang

Lotus Form 69

Shu Hua Zi

Hong Yun Fei Pian

Si He Lian

cv. Shu Hua Zi
Lotus form. Flowers 17cm × 3cm, purplish red (64-B); petals 3-whorled, dark purple basal blotches; stamens normal; pistils normal; floral discs yellowish white. Stalks stiff, flowers upright. Flowering early.

Plant medium height, spreading. Branches stiff, slightly curved. Leaves orbicular, medium-sized, stiff; leaflets ovate, obtuse at the apex. Growth vigorous, flowers many, tolerant of adverse conditions. Bred by Heze Flower Garden in 1994.

cv. Hong Yun Fei Pian
Lotus form. Flowers 19cm × 4cm, purplish red (61-C); petals broadly large, flat, purple basal blotches; stamens normal; pistils 3-5, stigmas purplish red; floral discs purplish red. Stalks stout, straight, flowers upright. Flowering early.

Plant tall, erect. Branches stout. Leaves long, large-sized; leaflets long elliptic, apex acute, pendulous, the margins slightly curved upwards; terminal leaflets deeply cut, veins sunken above. Growth vigorous, flowers many. Bred by Jingshan Park, Beijing during 1970s.

cv. Si He Lian
Lotus form. Flowers 14cm × 5cm, pink purplish red (68-A); petals broad and large, wavily wrinkled, purple basal blotches; stamens normal; pistils normal, very fertile. Stalks long, straight. Flowering early.

Plant tall, erect. Branches slender, stiff. Leaves long, medium-sized; leaflets long ovate, acuminate at the apex. Growth fairly vigorous, flowers many. Classic variety.

Hong Ling Yan

cv. Hong Ling Yan

Lotus form. Flowers 16cm × 4cm, light red (52-C); petals 3-4-whorled, broadly ovate, the margins slightly rugose, purple basal blotches; stamens normal, occasionally petaloid; pistils normal, stigmas purplish red; floral discs purplish red. Stalks stout, straight. Flowering midseason.

Plant dwarf, partially spreading. Branches stout. Leaves long, small-sized; leaflets lanceolate or elliptic, acuminate at the apex, the margins curved upwards. Growth fairly vigorous, flowers many, numerous branching stems. Bred by Wangcheng Park, Luoyang in 1995.

cv. Qin Hong

Lotus form, sometimes anemone form occurs. Flowers 12cm × 4cm, red (52-A); petals thin, soft, arranged sparsely, deeply toothed at the apex, dark purple basal blotches; stamens slightly petaloid; pistils small, stalks slender, pliable, flowers lateral. Flowering midseason.

Plant dwarf, spreading. Branches slender, pliable. Leaves long, small-sized, sparse; leaflets long-ovate or ovate lanceolate, apex acuminate, the margins curved upwards. Growth weak, flowers many. Classic variety.

Qin Hong

Xiang Yang Hong

Zi Yu Lan

Chun Hong Zheng Yan

cv. Zi Yu Lan

Lotus form. Flowers 16cm × 6cm, pinkish purple (68-C), pinkish white at the apex when fully open; petals 3-4-whorled, large, shallowly toothed, suffused with deep purple at the base; stamens normal; pistils normal; floral discs purplish red. Stalks fairly slender, pliable, flowers lateral. Flowering early.

Plant medium height, spreading. Branches fairly slender. Leaves long, small-sized, crowded; leaflets elliptic, apex acute, the margins curved upwards. Growth medium, flowers many. Bred by Wangcheng Park, Luoyang in 1969.

cv. Chun Hong Zheng Yan

Lotus form. Flowers 13cm × 3cm, light red (68-A); petals stiff, apex toothed, suffused with purplish red at the base; stamens normal; pistils normal; floral discs purplish red. Stalks long, stiff, flowers upright. Flowering early.

Plant dwarf, erect. Branches slender, stiff. Leaves long, small-sized, stiff, sparse; leaflets long ovate, apex acuminate. Growth dwarf, flowers many. Classic variety.

cv. Xiang Yang Hong

Lotus form. Flowers 18cm × 4cm, deep purplish red (61-B), lustrous; petals 4-whorled, large, entire and regular, lustrous, tolerant of strong sunshine, with a deep-cut at the apex, dark purple basal blotches; stamens normal; pistils normal; floral discs pinkish white. Stalks slender, long, stiff, straight, flowers upright. Flowering early.

Plant tall, erect. Branches slender, stiff, suffused with purple near the petiole axils. Leaves long, medium-sized, fairly sparse; leaflets ovate or long ovate, the margins curved upwards, slightly suffused with purple, acuminate at the apex, with conspicuous veins, light green. Growth vigorous, flowers many. Bred by Zhaolou Peony Garden, Heze in 1988.

cv. Zi Xia Ling

Lotus form. Flowers 16cm × 5cm, purplish red (53-B); petals fairly thin, apex toothed; stamens normal; pistils normal, stigmas purplish red, very fertile; floral discs deep purple. Stalks relatively slender, flowers upright. Flowering early.

Plant dwarf, partially spreading. Branches fairly slender, stiff. Leaves long, medium-sized, fairly crowded; leaflets long ovate or long elliptic, acuminate at the apex, the margins slightly curved upwards. Growth fairly vigorous. Bred by Zhaolou Peony Garden, Heze in 1968.

cv. Zhu Sha Lei

Lotus form. Flowers 17cm × 3cm, light red with slightly purple (73-A); petals broad and large, nearly regular and entire, suffused with purple at the base; stamens occasionally petaloid; pistils normal, very fertile. Stalks short, flowers lateral. Flowering midseason.

Plant medium height, partially spreading. Branches stout. Leaves orbicular, large-sized, fairly thick; leaflets ovate, deeply lobed, apex obtuse, pendulous. Growth vigorous, flowers many. Classic variety.

cv. Po Mo Zi

Lotus form. Flowers 11cm × 4cm, purplish red (53-B); petals, dark purple basal blotches; stamens normal; pistils normal, very fertile; floral discs dark purple. Stalks slender, straight, stiff, flowers lateral. Flowering late midseason.

Plant dwarf, partially spreading. Branches slender. Leaves long, small-sized, thin, stiff; leaflets ovate elliptic, obtuse at the apex, upper surface deep green, with light yellowish green blotches. Growth weak, flowers many. Classic variety.

cv. Zao Chun Hong

Lotus form. Flowers 15cm × 6cm, light red (47-D); petals 3-4-whorled, ovate, with translucent streaks, purple basal blotches; a few stamens petaloid; pistils small or petaloid. Stalks straight, flowers upright. Flowering midseason.

Plant tall, spreading. Branches slender. Leaves long, large-sized; leaflets long elliptic or ovate, thick, slightly curved upwards, acute at the apex. Growth medium, flowers many. Bred by Luoyang Peony Garden, Luoyang in 1995.

Zi Xia Ling

Zhu Sha Lei

Po Mo Zi

Lotus Form 73

Zao Chun Hong

Da Tao Hong

Shen Zi Yu

Ling Hua Xiao Cui

Luo Han Hong

Gong E Jiao Zhuang

cv. Da Tao Hong

Lotus form. Flowers 16cm × 4cm, deep pinkish red (52-D); petals 3-4-whorled, stiff, slightly wrinkled, colour pure, flowering period long; stamens normal; pistils normal; floral discs deep purplish red. Stalks long, straight, flowers upright. Flowering early.

Plant tall, erect. Branches slender, stiff. Leaves long, small-sized; leaflets ovate or long ovate, lateral leaflets lanceolate, acuminate at the apex. Growth vigorous, flowers many; tolerant of adverse conditions. Bred by Heze Flower Garden in 1989.

cv. Shen Zi Yu

Lotus form. Flowers 16cm × 4cm, deep purplish red (61-B); petals 4-whorled, the outer ones stiff and flat, the inner ones wrinkled, suffused with dark purple at the base; stamens normal, sometimes petaloid, pistils normal, stigmas pink; floral discs bluish pink. Stalks slightly short, flowers upright. Flowering early.

Plant medium height, partially spreading. Branches stout. Leaves long, medium-sized; the petiole axils suffused with purple; leaflets broadly ovate, terminal leaflets deeply cut, the margins slightly curved upwards. Growth vigorous, flowers many. Bred by Heze Flower Garden in 1973.

cv. Ling Hua Xiao Cui

Lotus form, sometimes anemone or crown form occurs. Flowers 18cm × 6cm, pale purple faintly pink (73-A); the outer petals soft, irregular, suffused with purplish red at the base; stamens normal or partially developed, petaloid, sparse, wrinkled; pistils small, floral discs white. Stalks long, pliable; flowers laterally pendulous. Flowering midseason.

Plant medium height, spreading. Branches slender, pliable, curved. Leaves long, medium-sized, thick, sparse; leaflets long ovate, obtuse at the apex, pendulous. Growth weak, flowers many. Classic variety.

cv. Luo Han Hong

Lotus form, sometimes crown form occurs. Flowers 16cm × 6cm, light red (52-C); petals stiff, suffused with purplish red at the base; stamens normal or some petaloid; pistils normal, occasionally infertile or petaloid. Stalks stiff. Flowering late midseason.

Plant dwarf, erect. Branches stout. Leaves orbicular, medium-sized; leaflets ovate with some lobes, the upper surface green suffused with purple. Growth vigorous, flowers few. Classic variety.

cv. Gong E Jiao Zhuang

Lotus form. Flowers 15cm × 3cm, red (52-C) with pink; petals 4-5-whorled, stiff, arranged regularly, of similar size, layered conspicuously, dark purple basal blotches; stamens normal; pistils normal; floral discs purplish red. Stalks slender, stiff, fairly long, flowers upright. Flowering early midseason.

Plant medium height, erect. Branches slender, stiff. Leaves long, medium-sized, stiff, sparse; leaflets ovate, apex acuminate. Growth vigorous; flowers many; tolerant of adverse conditions. Bred by Zhaolou Peony Garden, Heze in 1990.

Chrysanthemum Form

6 whorls. Petals gradually decreasing in size towards the centre. Stamens normal or fewer and petaloid in the centre of the flower; pistils normal.

Dan Yang

cv. Dan Yang

Chrysanthemum form. Flowers 15cm × 5cm, red (43-B); the outer petals 3-whorled, long ovate, broad and large, pale purple basal blotches; the inner petals wrinkled, stamens normal; pistils normal, floral discs purplish red. Stalks stout, straight; flowers blooming above leaves, upright. Flowering midseason.

Plant dwarf. Branchlets long, internodes fairly long. Leaves long, medium-sized; leaflets broadly lanceolate, lobes a few, acuminate at the apex, pendulous, upper surface smooth, yellowish green. Growth medium, flowers normal. Bred by Wangcheng Park, Luoyang in 1995.

cv. Bi Bo Xia Ying

Chrysanthemum form. Flowers 18cm × 5cm, bluish pink (62-C); petals 6-7-whorled, stiff, toothed at the apex, suffused with pale purple at the base; stamens normal; pistils 5-6; floral discs purplish red. Stalks long, stiff, flowers upright or lateral. Flowering late.

Plant medium height, partially spreading. Branches relatively stout. Leaves orbicular, medium-sized; petiolules long, slightly curved; leaflets ovate or elliptic, flat, acute at the apex. Growth vigorous, flowers profuse. Bred by Zhaolou Peony Garden, Heze in 1976.

Bi Bo Xia Ying

cv. Xu Ri

Chrysanthemum form. Flowers 15cm × 5cm, red (50-B); petals 8-whorled, the outer and the inner ones of the same colour; stamens normal, occasionally petaloid; pistils small or petaloid, floral discs purplish red. Stalks short, green, flowers upright. Flowering midseason.

Plant dwarf, partially spreading. Branches relatively stout, branchlets short, internodes short. Leaves long, small-sized, fairly crowded; leaflets ovate lanceolate, with a few and deepcut lobes, obtuse at the apex, the margins wavy upwards, the upper surface rough, deep green. Growth medium, flowers many. Bred by Zhaolou Peony Garden, Heze in 1983.

Xu Ri

cv. Xiu Yue

Chrysanthemum form. Flowers 16cm × 6cm, pinkish white (36-D); petals 6-8-whorled, broad and large, entire and regular, stamens normal, occasionally petaloid; pistils normal, stigmas purplish red; floral discs purplish red. Stalks stout, flowers lateral. Flowering midseason.

Plant medium height, spreading. Leaves long, medium-sized; leaflets ovate or long ovate, acuminate at the apex, pendulous. Growth medium, flowers many. Bred by Wangcheng Park, Luoyang in 1995.

Xiu Yue

cv. Jia Li

Chrysanthemum form. Flowers 14cm × 5cm, pinkish white faintly bluish (65-D); the outer petals 8-10-whorled, stiff, gradually shorter from the outer to the inner whorls, regular, purplish red basal blotches, fairly large; stamens normal; pistils 10-12, floral discs purple. Stalks long, stiff, flowers upright. Flowering midseason.

Plant tall, erect. Branches relatively slender, stiff, straight. Leaves long, small-sized, thin, sparse; leaflets long ovate, terminal leaflets deeply tri-lobed, acuminate at the apex. Growth vigorous; flowers profuse, flowering period long, tolerant of salinity-alkalinity. Bred by Zhaolou Peony Garden, Heze in 1970.

cv. Xiao Hu Die

Chrysanthemum form, occasionally in rose form. Flowers 16cm × 6cm, pink (62-C), the outer petals 4-5-whorled, stiff, regular, flat, layered conspicuously, with purple radial streaks, deep purplish red basal streaks; stamens smaller or petaloid; pistils normal, very fertile. Stalks long, flowers upright. Flowering midseason.

Plant medium height, erect. Branches slender and stiff. Leaves orbicular, medium-sized, stiff; leaflets elliptic, obtuse at the apex, the margins curved upwards, suffused with deep purple. Growth fairly vigorous, flowers many. Bred by Zhaolou Peony Garden, Heze in 1969.

cv. Ting Ting Yu Li

Chrysanthemum form. Flowers 15cm × 4cm, white (155-D); petals 5-whorled, thin, stiff, regular, pure white at the base; stamens normal, filaments white; pistils 7-8, stigmas red; floral discs white. Stalks slender, long, stiff, flowers upright. Flowering late midseason.

Plant tall, erect. Branches slender, stiff. Leaves long, small-sized, sparse; leaflets long ovate-lanceolate, acuminate at the apex, the margins slightly wavy. Growth medium; flowers many; tolerant of adverse conditions. Bred by Zhaolou Peony Garden, Heze in 1995.

Jia Li

Xiao Hu Die

Ting Ting Yu Li

Chrysanthemum Form

Da Hu Die

Ru Hua Si Yu

Man Jiang Hong

cv. Da Hu Die

Chrysanthemum form. Flowers 16cm × 4cm, deep pink bluish (62-B); the outer petals 5-6-whorled, thick, stiff, with a conspicuous purple streak along petals, purplish red basal blotches; stamens slightly petaloid; pistils 5-7, stigmas yellowish white, very fertile; floral discs yellowish white. Stalks long, straight, flowers upright. Flowering midseason.

Plant tall, erect. Branches slender, stiff. Leaves orbicular, medium-sized, thick, stiff, sparse; leaflets ovate, obtuse at the apex, the margins curved upwards, suffused with purplish red. Growth fairly vigorous, flowers many. Bred by Zhaolou Peony Garden, Heze in 1969.

cv. Ru Hua Si Yu

Chrysanthemum form. Flowers 15cm × 4cm, pink bluish (73-C); petals multi-whorled, stiff, gradually shorter from the outer to the inner petals, arranged closely, lighter colour at the apex, suffused with purple at the base; stamens normal; pistils normal, floral discs purplish red. Stalks slender, stiff, long, flowers upright. Flowering late midseason.

Plant tall, erect. Branches slender, stiff. Leaves long, small-sized, sparse; leaflets ovate lanceolate or long ovate, acuminate at the apex. Growth vigorous; flowers profuse. Bred by Heze Flower Garden in 1985.

cv. Man Jiang Hong

Chrysanthemum form. Flower buds oblate; flowers 16cm × 5cm, red (57-C); petals 5-6-whorled, stiff, regular, with a few teeth at the apex, suffused with purplish red at the base, the inner petals slightly rugose; stamens normal; pistils normal; floral discs purplish red. Stalks stiff; flowers upright. Flowering midseason.

Plant dwarf, partially spreading. Branches stiff. Leaves long, medium-sized, stiff, crowded; leaflets long ovate, the margins slightly curved upwards, apex acuminate. Growth medium, flowers normal. Bred by Zhaolou Peony Garden, Heze in 1962.

Bao Gong Mian

Chun Hong Jiao Yan

Yu Pan Tuo Jin

cv. Bao Gong Mian

Chrysanthemum form, sometimes rose form occurs. Flowers 17cm × 6cm, dark purple (187-C); petals 8-whorled, gradually shorter from the outer to the inner whorls, soft, dark purple basal streaks; stamens partially petaloid; pistils small, floral discs dark purple. Stalks slender, stiff, flowers upright. Flowering late.

Plant dwarf, partially spreading. Branches relatively stout. Leaves orbicular, medium-sized, crowded; leaflets long ovate, apex acute, the margins wavy upwards. Growth medium, flowers many. Bred by Zhaolou Peony Garden, Heze in 1995.

cv. Chun Hong Jiao Yan

Chrysanthemum form. Flowers 20cm × 5cm, light red (44-D), soft; petals multi-whorled, stiff, gradually shorter from the outer to the inner whorls, regular, dark purple basal streaks; stamens normal, pistils 7-11, floral discs purplish red. Stalks long, stiff, flowers upright. Flowering midseason.

Plant medium height, erect. Branches stout. Leaves orbicular, large-sized, stiff, thick; leaflets ovate, obtuse at the apex, the margins slightly curved upwards. Growth vigorous; flowers many, bright-colour, neatform. Bred by Heze Flower Garden in 1970.

cv. Yu Pan Tuo Jin

Chrysanthemum form. Flowers 15cm × 4cm, white faintly pink at first (56-D); petals 6-whorled, arranged regularly, stiff, wrinkled on the margins, deep purplish red basal blotches; stamens normal, pistils 11-13, floral discs pale purple. Stalks long, stiff, flowers upright. Flowering late.

Plant tall, erect. Branches slender, stiff. Leaves long, medium-sized, sparse; leaflets long ovate, terminal leaflets divided completely, acuminate at the apex. Growth vigorous; flowers many. Bred by Heze Flower Garden in 1984.

cv. Bi Yue Xiu Hua

Chrysanthemum form. Flower buds oblate; flowers 16cm × 4cm, light red (52-C), soft; petals multi-whorled, regular, suffused with light red at the base; stamens normal; pistils normal; floral discs purplish red. Stalks slender, stiff, flowers upright. Flowering early midseason.

Plant medium height, erect. Branches relatively slender, stiff. Leaves orbicular, small-sized, thick, slightly crowded; leaflets ovate, acute at the apex, the margins slightly wavy, suffused with purple. Growth medium; flowers many. Bred by Heze Flower Garden in 1974.

cv. Li Hua Fen

Chrysanthemum form. Flower buds conical, usually opening at the top; flowers 17 cm × 5cm, pink bluish (62-C), pinkish white when fully open; petals multi-whorled, the margins ruffled, suffused with red at the base, arranged regularly; stamens normal; pistils 11-13. Stalks straight, stiff, flowers upright. Flowering midseason.

Plant medium height, partially spreading. Branches stout. Leaves orbicular, large-sized; leaflets ovate elliptic, apex acute, terminal leaflets slightly pendulous. Growth vigorous, flowers many. Bred by Zhaolou Peony Garden, Heze in 1967.

cv. Cong Zhong Xiao

Chrysanthemum form. Flowers 14cm × 5cm, light red, the apex pink (52-D) suffused with purplish red on the margins; petals 6-7-whorled, stiff, arranged regularly, dark purple basal blotches; stamens normal; pistils numerous, floral discs pale purple. Stalks short, flowers upright. Flowering midseason.

Plant medium height, erect. Branches relatively slender, stiff. Leaves long, small-sized, fairly crowded; leaflets ovate, deeply lobed, acuminate at the apex, suffused with purplish red on the margins. Growth vigorous, flowers many. Bred by Zhaolou Peony Garden, Heze in 1965.

cv. Wan Xia

Chrysanthemum form. Flowers 16cm × 5cm, red (57-D); petals stiff, regular; stamens normal; pistils normal, very fertile. Stalks slender, straight, long; flowers upright. Flowering late.

Plant tall, erect. Branches relatively stout. Leaves long, medium-sized, sparse; leaflets long ovate, acute at the apex, the margins conspicuously curved upwards, suffused with deep purple on the margins. Growth medium, flowers many. Bred by Zhaolou Peony Garden, Heze in 1969.

Bi Yue Xiu Hua

Li Hua Fen

Chrysanthemum Form

Cong Zhong Xiao

Wan Xia

cv. Di Yuan Chun

Chrysanthemum form. Flowers 18cm × 5cm, pinkish red (61-D); petals 6-7-whorled, the outer petals entire and regular, fairly large, the inner petals shorter, with large purple blotches at the base; stamens normal, occasionally petaloid, pistils normal; floral discs purplish red. Stalks straight, flowers upright. Flowering midseason.

Plant tall, erect. Branches stout. Leaves long, medium-sized, thin; leaflets large, ovate, acute at the apex. Growth vigorous, flowers many. Bred by Xiyuan Park, Luoyang in 1995.

cv. Lan Hai Bi Bo

Chrysanthemum form. Flowers 18cm × 5cm, pink slightly bluish (69-B); petals stiff, arranged regularly, layered conspicuously, shallowly toothed at the apex, dark purple basal streaks; stamens normal, pistils petaloid or infertile. Stalks stiff, flowers upright. Flowering midseason.

Plant medium height, partially spreading. Branches relatively stout. Leaves long, medium-sized, stiff, slightly crowded; leaflets long ovate, the margins slightly curved upwards. Growth medium, flowers relatively few, tolerant of strong sunshine. Bred by Zhaolou Peony Garden, Heze in 1966.

cv. Zhao Yang Hong

Chrysanthemum form, sometimes rose form occurs. Flowers 18cm × 5cm, light red (41-C); petals multi-whorled, the outer petals arranged regularly, suffused with red at the base; stamens slightly petaloid; pistils small; floral discs red. Stalks stout, stiff, flowers upright. Flowering midseason.

Plant medium height, partially spreading. Branches stout. Leaves long, medium-sized, stiff, fairly crowded; leaflets long ovate, acuminate at the apex. Growth medium; flowers many. Bred by Zhaolou Peony Garden, Heze in 1968.

cv. Man Yuan Chun

Chrysanthemum form. Flowers 17cm × 7cm, pinkish purple faintly bluish (73-B); petals 6-whorled, broad and large, entire and regular, obovate, gradually smaller from outer to inner whorls; stamens occasionally petaloid, pistils normal. Stalks stout, straight, flowers lateral. Flowering midseason.

Plant tall, spreading. Branches stout. Leaves long, large-sized; leaflets long elliptic, apex acute, the margins slightly curved downwards, thick. Growth vigorous, flowers many. Bred by Wangcheng Park, Luoyng in 1995.

cv. Ge Jin Zi

Chrysanthemum form, sometimes rose form occurs. Flowers 16cm × 5cm, purple (64-D); petals 5-6-whorled, thin, soft, wrinkled, suffused with deep purplish red at the base, stamens partially petaloid; pistils small. Stalks slender, long, pliable, flowers lateral. Flowering late.

Plant dwarf, partially spreading. Branches slender. Leaves orbicular, medium-sized, soft, thick, fairly crowded; leaflets ovate, acute at the apex. Growth weak, flowers relatively few. Classic variety.

Zhao Yang Hong

cv. Ming Yuan

Chrysanthemum form. Flowers 19cm × 6cm, deep pinkish red (58-C); petals 6-8-whorled, stiff, regular, the apex irregular and toothed; stamens normal; pistils 5, floral discs deep purple. Stalks long, stiff, flowers upright. Flowering early midseason.

Plant tall, erect. Branches slender, stiff. New branches growing about 30cm long in 20-year plants. Leaves long, medium-sized; leaflets long ovate or broadly lanceolate, acuminate at the apex, the margins slightly curved upwards. Growth vigorous; flowers profuse. Bred by Zhaolou Peony Garden, Heze in 1970.

cv. Qie Lan Dan Sha

Chrysanthemum form. Flowers 17cm × 5cm, deep purple (74-C); petals multi-whorled, fairly soft, thin, wrinkled, dark purple basal blotches, with numerous teeth at the apex; stamens slightly petaloid; pistils small; stalks slender, stiff, flowers upright. Flowering midseason.

Plant relatively tall, erect. Branches slender, stiff. Leaves long, small-sized, stiff, sparse; leaflets long ovate, deeply lobed, slightly curved downwards, terminal leaflets slightly curved downwards, upper surface yellowish green, lustrous, with sunken veins, suffused with pale purple. Growth fairly vigorous, flowers many. Classic variety.

Di Yuan Chun

Lan Hai Bi Bo

cv. Luo Nu Zhuang

Chrysanthemum form. Flowers 14cm × 6cm, pinkish red (68-C); petals ovate with translucent streaks; stamens normal, occasionally petaloid, pistils normal. Stalks stout, straight, flowers upright. Flowering midseason.

Plant medium height, compact, partially spreading. Branches stout. Leaves long, medium-sized; leaflets long ovate, acuminate at the apex, the margins slightly curved upwards, suffused with purple. Growth vigorous, flowers many. Bred by Wangcheng Park, Luoyang in 1995.

Man Yuan Chun

Ge Jin Zi

Ming Yuan

Chrysanthemum Form

Qie Lan Dan Sha

Luo Nü Zhuang

Yan Hong Jin Bo

Tao Yuan Chun Se

Hei Hua Kui

Luo Nü Ying

cv. Yan Hong Jin Bo
Chrysanthemum form. Flower buds oblate; flowers 17cm × 4cm, light red (46-D); petals multi-whorled, stiff, regular, purplish red basal blotches; stamens normal; pistils 6-9; floral discs deep purplish red. Stalks slender, stiff, dark purple, flowers upright. Flowering early.

Plant tall, partially spreading. Branches slender, stiff. Leaves long, small-sized, thin; leaflets long ovate, acuminate at the apex. Growth medium; flowers many, tolerant of low temperature during bud stage. Bred by Heze Flower Garden in 1972.

cv. Hei Hua Kui
Chrysanthemum form. Flowers 17cm × 6cm, dark purple (187-A); petals 6-8-whorled, full flowers rugose, suffused with dark purple at the base; stamens normal, occasionally petaloid; pistils normal or occasionally developed, petaloid, greenish coloured. Stalks slightly short, pliable, purplish brown, flowers lateral. Flowering midseason.

Plant dwarf, partially spreading. Branches relatively slender. Leaves orbicular, medium-sized, soft; petiolules slightly short, leaflets ovate, acute at the apex, upper surface smooth, suffused with pale purple. Growth weak, flowers many. Classic variety.

cv. Luo Nu Ying
Chrysanthemum form. Flowers 15cm × 6cm, pinkish red (49-A), lustrous; petals broad and large, regular, gradually shorter from the outer to the inner whorls, stamens normal or partially petaloid, pistils normal; floral discs deep purplish red. Stalks fairly pliable, flowers lateral. Flowering early.

Plant dwarf, partially spreading. Branches relatively stout. Leaves long, medium-sized, sparse; leaflets long elliptic, acute at the apex, slightly suffused with red on the margins, slightly curved upwards. Growth medium, flowers many. Bred by Wangcheng Park, Luoyang in 1969.

cv. Tao Yuan Chun Se
Chrysanthemum form. Flowers 14cm × 6cm, deep red (67-B); petals broadly ovate, purple basal blotches; stamens normal, occasionally petaloid; pistils normal; occasionally developed, petaloid, greenish coloured. Stalks stout, straight, flowers upright. Flowering midseason.

Plant dwarf. Branches stout. Leaves long, small-sized; leaflets lanceolate, acute at the apex. Growth vigorous, flowers many. Bred by Wangcheng Park, Luoyang in 1995.

cv. Dong Hai Lang Hua

Chrysanthemum form. Flowers 20cm × 6cm, bluish pink (62-D); petals 6-8-whorled, broad and large, ovate, gradually shorter from the outer to the inner whorls, suffused with purple at the base; stamens occasionally petaloid; pistils smaller, stigmas white; floral discs white. Stalks stout, straight, flowers lateral. Flowering midseason.

Plant tall, erect. Branches stout. Leaves long, large-sized; leaflets long elliptic, acute at the apex; upper surface smooth, with slightly purple. Growth vigorous, flowers average. Bred by Wangcheng Park, Luoyang in 1995.

Jing Yun

cv. Jing Yun

Chrysanthemum form. Flowers 20cm × 9cm, pale purplish red (67-C); petals obovate, broad and large, regular, purple basal blotches, the margins wrinkled; stamens normal; pistils normal, stigmas purplish red; floral discs purplish red. Stalks stout, straight, flowers upright. Flowering midseason.

Plant tall, erect. Branches stout. Leaves long, medium-sized; leaflets long elliptic, acuminate at the apex, the margins slightly curved upwards; terminal leaflets deeply cut; upper surface yellowish green, veins sunken. Growth vigorous, flowers many. Bred by Jingshan Park, Beijing during 1970s.

Chrysanthemum Form 87

Dong Hai Lang Hua

cv. Rou Fu Rong

Chrysanthemum form, sometimes proliferate form occurs. Flowers 17cm × 5cm, deep pink (52-D); petals thin, soft, rugose, suffused with purple at the base; stamens slightly petaloid; pistils small. Stalk fairly short, flowers upright or lateral. Flowering midseason.

Plant medium height, spreading. Leaves long, medium-sized; leaflets long ovate with long petiolules, acuminate at the apex, the margins slightly curved upwards, rarely suffused with pale purple. Growth vigorous, flowers many; tolerant of adverse conditions and salinity-alkalinity. Bred by Team 10, Zhaolou, Heze in 1975.

cv. Luo Shen

Chrysanthemum form. Flowers 18cm × 6cm, light magenta (50-B), lustrous; the outer petals 4-5-whorled, broad and large, entire and regular, the inner petals slightly smaller, purple basal blotches, stamens normal, filaments purplish red; pistils normal; floral discs deep purplish red. Stalks straight, flowers upright. Flowering midseason.

Plant tall, erect. Branches relatively stout. Leaves orbicular, medium-sized; leaflets ovate, obtuse at the apex. Growth vigorous, flowers many. Bred by Wangcheng Park, Luoyang in 1995.

cv. Yu Ji Yan Zhuang

Chrysanthemum form, sometimes lotus form occurs. Flowers 18cm × 5cm, red (43-B), bright and lustrous; petals 5-7-whorled, thin, soft, with rugose teeth at the apex; stamens normal; pistils normal, very fertile. Stalks slender and stiff, flowers upright. Flowering late midseason.

Plant medium to dwarf, partially spreading. Branches relatively slender. Leaves long, medium-sized; leaflets long ovate, acuminate at the apex. Growth medium, flowers many. Bred by Zhaolou Peony Garden, Heze in 1980.

Rou Fu Rong

Luo Shen

Yu Ji Yan Zhuang

cv. Su Jia Hong

Chrysanthemum form, sometimes rose form occurs. Flowers 13cm × 5cm, light red (50-B), soft; petals 5-6-whorled, thin, soft, with translucent streaks, wavy and wrinkled at the apex, dark purple basal blotches, stamens slightly petaloid, pistils developed, petaloid, yellowish green coloured. Stalks slender, slightly pliable, flowers lateral. Flowering midseason.

Plant dwarf, spreading. Branches slender, curved. Leaves long, small-sized, sparse; leaflets ovate lanceolate, acuminate at the apex, the margins slightly curved upwards. Growth weak; flowers many; adaptability poor. Classic variety.

cv. Yong Chun

Chrysanthemum form, sometimes proliferate form occurs. Flowers 21cm × 8cm, magenta (50-B); petals large, slightly wrinkled and wavy, toothed at the apex, stamens normal; pistils normal or occasionally petaloid. Stalks fairly long, slightly pliable, flowers lateral. Flowering midseason.

Plant tall, spreading. Branches stout, fairly pliable. Leaves long, large-sized, sparse; petiolules long; leaflets lanceolate, fairly large, flat, acuminate at the apex. Growth vigorous, flowers many. Bred by Guose Peony Garden, Luoyang in 1992.

cv. Qie Hua Zi

Chrysanthemum form, sometimes hundred proliferate form. Flowers 16cm × 5cm, purple (68-A); petals 6-8-whorled, stiff, layered conspicuously, suffused with deep colour at the base; stamens normal, occasionally petaloid; pistils normal, stigmas purplish red; floral discs purplish red. Stalks straight. Flowering midseason.

Plant tall, partially spreading. Branches stout. Leaves orbicular, large-sized; leaflets ovate, acute at the apex, suffused with purple on the margins. Growth vigorous, flowers many. Bred by Zhaolou Peony Garden, Heze in 1967.

Su Jia Hong

Yong Chun

Qie Hua Zi

Xiao Qing

Tao Hua Fei Xue

Xiu Li Hong

Chen Hui

cv. Xiao Qing

Chrysanthemum form. Flowers 17cm × 6cm, pinkish red (62-B) faintly bluish; petals 8-whorled, slightly stiff, outer whorls 1-3, lighter colour, colour gradually deep from the outer to the inner petals, with deep purplish red streaks at the base; stamens normal, occasionally petaloid; pistils 11-13, small, slightly petaloid, floral discs purplish red. Stalks long, stiff, flowers upright. Flowering early midseason.

Plant medium height, erect. Branches slender, stiff. Leaves long, medium-sized, thin, sparse, stiff; leaflets broadly ovate, acuminate at the apex, slightly suffused with purple on the margins. Growth vigorous, flowers many, tolerant of adverse conditions, very adaptable. Bred by Zhaolou Peony Garden, Heze in 1996.

cv. Tao Hua Fei Xue

Chrysanthemum form. Flowers 18cm × 6cm, pink suffused with bluish (62-A); petals 8 whorled, stiff, regular; stamens normal or occasionally petaloid; pistils normal or small. Stalks fairly long, flowers upright. Flowering midseason.

Plant medium height, partially spreading. Branches fairly stout. Leaves long, medium-sized, fairly crowded, stiff; leaflets long ovate, acuminate at the apex, the margins curved upwards, suffused with pale purple. Growth medium, flowers many. Bred by Zhaolou Peony Garden, Heze in 1983.

cv. Chen Hui

Chrysanthemum form. Flowers 18cm × 6cm, light red (50-B), soft; petals multi-whorled, slightly stiff, regular, crowded, gradually smaller from the outer to the inner whorls, lighter colour at the apex, deep purplish red basal blotches, stamens normal, occasionally petaloid; pistils small; floral discs purple. Stalks short, stiff, flowers upright or lateral. Flowering midseason.

Plant dwarf, partially spreading. Branches stiff. Leaves orbicular, medium-sized, stiff, crowded; leaflets ovate, stiff, acute at the apex, the margins slightly curved upwards. Growth medium; flowers many. Bred by Heze Flower Garden in 1984.

cv. Xiu Li Hong

Chrysanthemum form. Flowers 15cm × 4cm, purplish red (53-B), lustrous; petals multi-whorled, arranged regularly, thick, stiff, dark purple basal blotches; stamens slightly petaloid; pistils 7-9, floral discs deep purple. Stalks fairly long, flowers upright. Flowering midseason.

Plant medium to dwarf, partially spreading. Branches relatively stout. Leaves long, medium-sized, thick, stiff; leaflets elliptic, acuminate at the apex, upper surface deep green. Growth fairly vigorous, flowers many. Bred by Zhaolou Peony Garden, Heze in 1967.

Chrysanthemum Form

Yan Zi Xian Jin

Zi Yan

He Ding Hong

Mo Su

cv. Yan Zi Xian Jin
Chrysanthemum form. Flowers 18cm × 6cm, purplish red (60-B), lustrous; petals suffused with dark purple at the base, stamens normal, filaments purplish red; pistils normal, stigmas purplish red; floral discs deep purple. Stalks straight, flowers slightly lateral. Flowering midseason.

Plant tall, erect. Branches relatively slender, stiff. Leaves orbicular, small-sized, thick; petioles 9cm long, green, horizontal; leaflets ovate, lateral leaflets pendulous at the tip. Growth vigorous, flowers many. Bred by Luoyang Peony Garden, Luoyang in 1995.

cv. He Ding Hong
Chrysanthemum form, sometimes rose form occurs. Flowers 17cm × 5cm, light red faintly purple (68-B); petals multi-whorled, arranged regularly, deep purple at the base; stamens occasionally petaloid, pistils numerous, small. Stalks stiff, straight, flowering upright. Flowering midseason.

Plant medium height, partially spreading. Branches relatively stout. Leaves long, medium-sized, crowded; leaflets long ovate or elliptic, acute at the apex. Growth vigorous, flowers many; resistant to diseases. Bred by Zhaolou Peony Garden, Heze in 1963.

cv. Zi Yan
Chrysanthemum form. Flowers 18cm × 5cm, pinkish purple with bluish (68-A); petals 6-8-whorled, the margins toothed, suffused with purplish red at the base, floral discs red, stamens normal; pistils normal, fertile. Stalks long, stiff, flowers upright. Flowering midseason.

Plant medium height, erect. Branches relatively stout. Leaves long, large-sized; leaflets long ovate or broadly ovate lanceolate, acuminate at the apex, curved downwards, the margins slightly curved upwards, dark green above, suffused with red. Growth fairly weak, flowers many. Bred by Zhaolou Peony Garden, Heze in 1967.

cv. Mo Su
Chrysanthemum form. Flowers 18cm × 5cm, dark purple (187-D), lustrous; petals thick, stiff, slightly rugose, arranged regularly, suffused with black at the base; stamens normal, occasionally petaloid; pistils 9-11, nearly infertile; stalks long, straight, flowers upright. Flowering late midseason.

Plant medium height, erect. Branches stout. Leaves long, medium-sized, thin, stiff; leaflets long ovate with numerous lobes, acute at the apex, with conspicuous veins, upper surface green. Growth fairly vigorous; flowers many. Bred by Team 6, Zhaolou, Heze in 1974.

Bai Hua Cong Xiao

Bai Hua Cong Xiao

cv. Bai Hua Cong Xiao

Chrysanthemum form. Flowers 18cm × 6cm, light red (64-C) slightly purple; petals multi-whorled, stiff, slightly wrinkled, regular, crowded, paler purple at the base, stamens normal, pistils 9-11, occasionally petaloid, floral discs pale purple. Stalks long, stiff, flowers upright. Flowering late midseason.

Plant tall, erect. Branches stiff. Leaves long, medium-sized, stiff, sparse; leaflets elliptic or ovate lanceolate, acuminate at the apex, pendulous. Growth vigorous; flowers many; tolerant of low temperature during bud stage. Bred by Heze Flower Garden in 1978.

94 Chinese Tree Peony

Yang Guang

Hong Yun

Tao Yuan Xian Jing

Ceng Zhong Xiao

Ao Yang

Wan Xia Yu Hui

Dou Kou Nian Hua

cv. Yang Guang
Chrysanthemum form. Flowers 20cm × 9cm, light magenta (55-A), lustrous; petals multi-whorled, gra-dually shorter towards the centre, nearly entire and regular, deep purplish red basal blotches, stamens normal; pistils normal, stigmas deep purplish red; floral discs greenish white. Stalks fairly pliable, flowers lateral. Flowering midseason.
Plant medium height, spreading. Branches slender, pliable. Leaves long, medium-sized, sparse; leaflets long ovate, acuminate at the apex. Growth medium, flowers normal. Bred by Guose Peony Garden, Luoyang in 1992.

cv. Tao Yuan Xian Jing
Chrysanthemum form. Flowers 20cm × 4cm, deep pinkish red (52-D); petals 8-10-whorled, stiff, gradually shorter from outer to the inner petals, arranged regularly, layered conspi-cuously; the inner petals often cut, light in colour and pinkish white at the base; stamens normal; pistils normal; slightly petaloid, floral discs purple. Stalks short, straight, flowers lateral, slightly hidden among leaves. Flowering early.
Plant medium height, partially spreading. Branches stout. Leaves long, medium-sized, sparse, soft; leaflets broadly-ovate, acute at the apex, the margins slightly curved. Growth vigorous, flowers many; tolerant of adverse conditions. Bred by Zhaolou Peony Garden, Heze in 1986.

cv. Ao Yang
Chrysanthemum form. Flowers 16cm × 4cm, purplish red (53-C), lustrous; petals, stiff, regular, conspi-cuously layered; stamens normal, small; floral discs purple. Stalks slender, stiff, flowers upright. Flowering midseason.
Plant medium height, erect. Branches fairly slender, stiff. Leaves long, medium-sized, stiff; leaflets long ovate, acuminate at the apex. Growth vigorous but sensitive to bad weather during early spring; young flower buds often abort or flower irregularly due to temperature change, which may reduced rate of flowering; flowers relatively few. Bred by Zhaolou Peony Garden, Heze in 1963.

cv. Hong Yun
Chrysanthemum form. Flowers 16cm × 4cm, pale purplish red (64-C); petals 5-6-whorled, regular, slightly wrinkled and toothed at the apex, suffused with purplish red at the base, stamens normal; pistils normal, floral discs purplish red. Stalks slender, stiff, flowers upright. Flowering late.
Plant tall, partially spreading. Branches relatively stout. Leaves long, small-sized, thick, sparse; leaflets ovate-lanceolate, acuminate at the apex. Growth vigorous, flowers many; tolerant of strong sunshine and of low temperature during flower bud stage. Bred by Heze Flower Garden in1992.

cv. Ceng Zhong Xiao
Chrysanthemum form. Flowers 16cm × 6cm, pale purplish red (67-C); petals multi-whorled, the outer ones flat, the inner ones fairly small, lighter colour at the apex, with numerous teeth, stamens normal; pistils normal; floral discs purplish red. Stalks slender, long, flowers upright. Flowering midseason.
Plant tall, erect. Branches slender, stiff. Leaves long, medium-sized, sparse; leaflets ovate lanceolate, acuminate at the apex, slightly purple on the margins. Growth vigorous, flowers many. Bred by Heze Flower Garden in 1995.

cv. Wan Xia Yu Hui
Chrysanthemum form. Flowers 20cm × 5cm, purplish red (60-A); petals 8-whorled, the outer 2-3-whorls, large, toothed at the apex, the inner petals conspicuously smaller; stamens normal; pistils 5, floral discs purplish red. Stalks fairly long, stiff, flowers upright. Flowering midseason.
Plant tall, erect. Branches stout. Leaves long, small-sized; lateral leaflets ovate lanceolate, apex acute, upper surface yellowish green, slightly suffused with purple. Growth vigorous, flowers many; tolerant of spring frost. Bred by Zhaolou Peony Garden, Heze in 1986.

cv. Dou Kou Nian Hua
Chrysanthemum form. Flowers 14cm × 7cm, pinkish white faintly bluish (56-D); petals narrowly long, often shallowly toothed, regular, conspicuously with purplish red streaks at the base; stamens normal; pistils 9-13, normal, stigmas red; floral discs white. Stalks straight, flowers upright. Flowering midseason.
Plant medium height, erect. Branches relatively stout. Petiolules short; leaflets long elliptic, acute at the apex, the margins curved downwards. Growth vigorous, flowers many. Bred by Guose Peony Garden, Luoyang in 1992.

cv. Zi Yang

Chrysanthemum form. Flowers 15cm × 5cm, deep purplish red (61-B); petals ovate, suffused with deep purplish red at the base; stamens normal; pistils normal; occasionally petaloid, stigmas purplish red; floral discs purplish red. Stalks relatively stout. Flowering early.

Plant medium height, partially spreading. Branches relatively slender. Leaves long, small-sized; leaflets long elliptic, acute at the apex; upper surface rough, deep green. Growth medium, flowers many. Bred by Luoyang Peony Garden, Luoyang in 1995.

cv. Bai Tain E

Chrysanthemum form. Flowers 17cm × 4cm, white (155-D), pure white without blotches; petals multi-whorled, thin, soft, wrinkled, raised, the margins pinnately cut, white at the base; stamens normal; pistils normal, slightly petaloid; floral discs pale purple. Stalks stout, stiff, fairly long, flowers upright. Flowering early.

Plant medium height, erect. Branches slender, stiff. Leaves long, medium-sized, sparse, slightly soft; leaflets long ovate, broadly lanceolate, terminal leaflets deeply cut, twisted, apex acute, the margins suffused with pale purple. Growth vigorous, flowers many, very adaptable. Bred by Zhaolou Peony Garden, Heze in 1990.

cv. Feng Hua Zi

Chrysanthemum form. Flowers 15cm × 4cm, pale purplish red (67-B); petals ovate, entire and regular, with translucent streaks, purple basal blotches; stamens normal; pistils normal, stigmas purplish red; floral discs purplish red. Stalks stout, straight, flowers upright. Flowering midseason.

Plant tall, partially spreading. Leaves orbicular, medium-sized; leaflets ovate, apex acute, slightly suffused with purple on the margins; upper surface smooth, green. Growth vigorous, flowers many. Bred by Wangcheng Park, Luoyang in 1995.

cv. Yan Yi Xiang Rong

Chrysanthemum form. Flowers 15cm × 5cm, deep red (52-A), soft; petals 5-8-whorled, slightly stiff, arranged regularly, layered conspi-cuously, with red radical streaks; stamens normal; pistils normal, slightly petaloid. Stalks slender, stiff, flowers upright or lateral. Flowering early midseason.

Plant dwarf, partially spreading. Branches short, slender, stiff. Leaves orbicular, medium-sized, thin, stiff, relatively crowded; leaflets ovate or nearly ovate, apex acute, suffused with purple on the margins; rugose, light green above. Growth medium, flowers profuse. Bred by Zhaolou Peony Garden, Heze in 1990.

cv. Da Hong Yi Pin

Chrysanthemum form. Flowers 16cm × 4cm, deep red (61-A); petals 6-8-whorled, fairly stiff, entire and regular, purplish red basal blotches; stamens normal; pistils normal; floral discs purplish red. Stalks stout, stiff, short, flowers lateral. Flowering early.

Plant medium height, partially spreading. Branches stout. Leaves long, medium-sized, stiff, crowded; leaflets broadly-ovate or ovate, terminal leaflets deeply or completely cut, acuminate at the apex, the margins curved upwards, suffused with pale purple. Growth vigorous, flowers many, very adaptable. Bred by Zhaolou Peony Garden, Heze in 1990.

Zi Yang

Bai Tian E

Feng Hua Zi

Yan Yi Xiang Rong

Chrysanthemum Form 97

Da Hong Yi Pin

Shou Xing Hong

Pan Tao Zhou

Shou Xing Hong

cv. Pan Tao Zhou

Chrysanthemum form. Flowers 16cm × 7cm, light red (68-B); petals broad and large, suffused with purplish red at the base, gradually shorter from outer to inner whorls, rugose; stamens normal; pistils normal, occasionally petaloid. Stalks stout, straight. Flowering midseason.

Plant medium height, erect. Leaves long, medium-sized; leaflets long elliptic, acuminate at the apex. Growth vigorous, flowers normal. Bred by Wangcheng Park, Luoyang in 1995.

cv. Shou Xing Hong

Chrysanthemum form. Flowers 16cm × 6cm, purplish red (61-A), lustrous; petals 5-6-whorled, the apex shallowly toothed, suffused with dark purple at the base; stamens slightly petaloid; pistils numerous, small; floral discs deep purple. Stalks fairly long, flowers upright. Flowering midseason.

Plant medium height, partially spreading. Branches fairly stout. Leaves long, medium-sized, stiff, fairly crowded; leaflets ovate, acuminate at the apex, the margins curved upwards, suffused with deep purplish brown. Growth fairly vigorous, flowers many, leaves falling early. Bred by Zhaolou Peony Garden, Heze in 1967.

Chrysanthemum Form 99

Chun Se Man Yuan

Chun Se Man Yuan

Hong Yan Yan

cv. Chun Se Man Yuan

Chrysanthemum form. Flowers 16cm × 4cm, pale purple (62-B); petals multi-whorled, stiff, regular, the apex pinkish white, often toothed, purplish red basal blotches; stamens normal, pistils 7-11; floral discs purple. Stalks long, stiff; flowers upright. Flowering midseason.

Plant tall, erect. Branches stiff, slender. Leaves long, medium-sized, stiff; leaflets elliptic or ovate lanceolate, apex acuminate, the margins slightly wavy, upper surface rough, yellowish green suffused with purplish red. Growth vigorous; flowers profuse, tolerant of strong sunshine. Bred by Heze Flower Garden in 1994.

cv. Hong Yan Yan

Chrysanthemum form. Flowers 16cm × 5cm, purplish red (61-C); petals 6-8-whorled, obovate, purple basal blotches; stamens normal, occasionally petaloid, pistils normal, stigmas red; floral discs purplish red. Stalks stout, straight, flowers upright. Flowering midseason.

Plant tall, erect. Branches stout. Leaves long, medium-sized; leaflets orbicular or lanceolates, acute at the apex, the margins curved upwards; terminal leaflets deeply cut, pilose beneath. Growing vigorous, flowers many. Bred by Jingshan Park, Beijing during 1970s.

Hua Yuan Hong

cv. Hua Yuan Hong

Chrysanthemum form. Flowers 20cm × 8cm, bright purplish red (60-B); petals 5-whorled, stiff, twisted inward, suffused with deep purple at the base; stamens normal; pistils 5; floral discs pinkish purple. Stalks long, stiff, pale purple, flowers upright. Flowering late midseason.

Plant tall, erect. Branches stiff. Leaves long, large-sized, soft, slightly thick, crowded; leaflets elliptic, apex acuminate, upper surface deep yellowish green suffused with pale purple. Growth vigorous; flowers many, tolerant of adverse conditions. Bred by Heze Flower Garden in 1994.

cv. Hong Fu

Chrysanthemum form. Flowers 15cm × 6 cm, purplish red (53-B), bright, lustrous; petals 8-whorled, regular, stiff, suffused with dark purple at the base; stamens normal or occasionally petaloid; pistils 7-11; floral discs purplish red. Stalks fairly short, flowers lateral. Flowering midseason.

Plant tall, partially spreading. Branches relatively slender. Leaves long, medium-sized, soft; leaflets long ovate, apex acuminate, pendulous; upper surface green, with conspicuous veins. Growth vigorous, flowers many. Bred by Zhaolou Peony Garden, Heze in 1968.

Hua Yuan Hong

Hong Fu

Chrysanthemum Form

Yang Hong Ning Hui

Feng Ye Hong

Jin Gong Pao

cv. Yang Hong Ning Hui

Chrysanthemum form, Flowers 16cm × 7cm, red (58-B); petals 5-6-whorled, broad and large, ovate, gradually smaller from the outer to the inner whorls, base suffused with purplish red; stamens normal, occasionally petaloid; pistils normal. Stalks stout, straight. Flowering midseason.

Plant medium height, partially spreading. Leaves long, small-sized; leaflets elliptic, acute at the apex. Growth vigorous, flowers average. Bred by Wangcheng Park, Luoyang in 1995.

cv. Feng Ye Hong

Chrysanthemum form. Flowers 20cm × 5cm, pale purplish red (72-D); petals multi-whorled, large, stiff, the margins wavy; stamens normal; pistils 5-7; floral discs purple. Stalks stout, long, straight, flowers upright. Flowering late.

Plant tall, erect. Branches fairly stout. Leaves long, medium-sized, sparse; leaflets ovate lanceolate, acuminate at the apex. Growth vigorous, flowers many, tolerant of the spring frost. Bred by Zhaolou Peony Garden, Heze in 1988.

cv. Jin Gong Pao

Chrysanthemum form. Flowers 18cm × 6cm, purplish red (61-C); petals 8-whorled, wrinkled and toothed at the apex, curved near the centre; stamens normal; pistils small; floral discs purplish red. Stalks stiff, flowers upright. Flowering midseason.

Plant tall, partially spreading. Branches relatively stout. Leaves long, medium-sized; petiolules long, oblique; leaflets ovate lanceolate; terminal leaflets completely cut, lobes lanceolate; lateral leaflets with numerous and deep lobes; acute at the apex. Growth vigorous, flowers many, tolerant of the spring frost. Bred in Heze in 1981.

cv. Hong Mei Gui

Chrysanthemum form. Flowers 16cm × 5cm, purplish red (53-B); petals 6-8-whorled, the margins shallowly toothed, suffused with deep purple at the base; stamens normal; pistils normal; floral discs purplish red. Stalks stiff, flowers upright. Flowering midseason.

Plant medium height, erect. Branches relatively stout; branchlets long, internodes long. Leaves long, medium-sized, crowded; leaflets long elliptic, apex acuminate with some lobes, upper surface green. Growth vigorous, flowers many. Bred by Heze Flower Garden in 1995.

cv. Tao Hua Hong

Chrysanthemum form. Flowers 15cm × 6cm, pinkish purple (68-A); petals multi-whorled, purple basal blotches; the inner petals 1-2-whorled, very small, slightly rugose; stamens normal; pistils normal, stigmas red; floral discs purplish red. Stalks straight, flowers upright. Flowering midseason.

Plant tall, erect. Branches stout. Leaves orbicular, large-sized; petioles 14cm long, horizontal, purplish red; leaflets ovate, obtuse at the apex, slightly curved downwards, upper surface smooth, yellowish green. Growth vigorous, flowers many. Bred by Jingshan Park, Beijing.

Zi Xia Ying Jin

cv. Xian Gu

Chrysanthemum form. Flowers 20cm × 7cm, pale purplish red (63-B); petals 6-8-whorled, regular, suffused with deep purple at the base; stamens normal; pistils 6-9; floral discs purplish red. Stalks stiff, long, flowers upright. Flowering midseason.

Plant tall, erect. Branches relatively slender, stiff, branchlets long, internodes long. Leaves long, medium-sized; leaflets long ovate, acuminate at the apex, upper surface green. Growth vigorous, flowers many. Bred by Zhaolou Peony Garden, Heze in 1980.

cv. Zi Xia Ying Jin

Chrysanthemum form. Flowers 17cm × 5cm, pale purple (63-B); petals 7-8-whorled, broad and large, with streaks of deeper colour along the petal centre; the inner petals relatively small, some stamens petaloid; pistils normal, stigmas purplish red. Flowering midseason.

Plant medium height, partially spreading. Leaves long, small-sized; leaflets long elliptic, acute at the apex, suffused with pale purplish red on the margins. Growth vigorous, flowers many. Bred by Wangcheng Park, Luoyang in 1995.

Xian Gu

Rose Form

Petals generally longer than in chrysanthemum form, becoming smaller from the outside towards the centre. Most stamens disappeared. Pistils normal, slightly petaloid, reduced or completely disappeared.

Er Qiao

Rose Form 105

Da Zong Zi

Yu Hou Feng Guang

cv. Er Qiao

Rose form. Flowers 16cm × 6cm; flowers double coloured, with purplish red (61-C) and pink (38-D) on the same plant or on the same branch and flower; petals stiff, regular, dark purple basal blotches; stamens slightly petaloid; pistils 9-11, floral discs purple. Stalks long, stiff, flowers upright. Flowering midseason.

Plant tall, erect. Branches fairly slender, stiff. Leaves orbicular, medium-sized, stiff, slightly sparse; leaflets ovate, acuminate at the apex; leaflets with numerous shallowly-cut lobes growing with pink flowers; the leaflets with a few deep-cut lobes growing with purplish red flowers; leaves with two shapes appearing on the branches which develop with the double coloured flowers. Growth vigorous, flowers many. Classic variety.

cv. Da Zong Zi

Rose form. Flowers 17cm × 7cm, purplish red (61-B); petals multi-whorled, entire, arranged regularly, suffused with dark purple at the base; Stamens occasionally petaloid; pistils 9-11, infertile. Stalks long, straight, flowers upright. Flowering midseason.

Plant medium height, erect. Branches stout. Leaves orbicular, medium-sized, fairly thick; leaflets long ovate, obtuse at the apex, the margins slightly curved upwards, terminal leaflets in central leaflets completely tri-sected. Growth vigorous, flowers many. Classic variety.

cv. Yu Hou Feng Guang

Rose form. Flowers 19cm × 7cm, pink faintly bluish (73-C); the outer petals 3-whorled, large, purplish red basal blotches; the inner petals soft, wrinkled, apex toothed, crowded; pistils small or sometimes developed, petaloid, greenish coloured. Stalks long, slightly pliable, flowers lateral. Flowering midseason.

Plant tall, erect. Branches fairly slender, stiff. Leaves long, medium-sized; leaflets ovate with numerous lobes, apex acuminate, upper surface smooth, yellowish green. Growth vigorous; flowers many. Bred by Zhaolou Peony Garden, Heze in 1971.

Yi Pin Hong

He Yuan Hong

Ying Jin Hong

Luo Yang Hong

cv. Yi Pin Hong

Rose form, sometimes chrysanthemum form occurs. Flowers 17cm × 7cm, red (43-C); petals multi-whorled, apex toothed, purplish red basal blotches; stamens slightly petaloid; pistils small or petaloid, floral discs deep purplish red. Stalks short, stiff, flowers upright. Flowering midseason.

Plant medium height, partially spreading. Branches stout. Leaves orbicular, medium-sized, thick, crowded; leaflets broadly-ovate, apex acute. Growth medium, flowers many; tolerant of leaf spot and low temperature during bud stage. Bred by Heze Flower Garden in 1972.

cv. Ying Jin Hong

Rose form, sometimes chrysanthemum form occurs. Flowers 16cm × 6cm, deep purplish red (61-B). Petals thin, soft, with numerous teeth on the margins, suffused with dark purple at the base; stamens partially petaloid, pistils small. Stalks long, flowers lateral. Flowering midseason.

Plant tall, erect. Branches fairly slender. Leaves long, medium-sized, thin, stiff; leaflets ovate, acuminate at the apex, slightly pendulous, suffused with pale purple on the margins. Growth vigorous, flowers many, very adaptable. Bred by Zhaolou Peony Garden, Heze in 1963.

cv. He Yuan Hong

Rose form. Flowers 14cm × 5cm, red (52-A); the outer petals soft, reflexed, dark purple at the base; the inner petals wrinkled, soft, sparse; stamens partially petaloid, pistils small or petaloid. Stalks short, flowers upright. Flowering midseason.

Plant dwarf, partially spreading. Branches relatively slender. Leaves orbicular, medium-sized, relatively crowded; leaflets ovate or elliptic, obtuse at the apex, terminal leaflets soft, pendulous, upper surface deep green. Growth weak, flowers relatively few. Classic variety.

cv. Luo Yang Hong

Rose form, sometimes chrysanthemum form occurs. Flowers 16cm × 6cm, purplish red (61-C), lustrous; petals multi-whorled, stiff, arranged regularly, dark purple basal blotches; stamens partially petaloid; pistils numerous, small, sometimes fertile; floral discs dark purplish red. Stalks fairly long, stiff, flowers upright. Flowering midseason.

Plant tall, erect. Branches relatively slender, stiff. Leaves long, medium-sized, stiff; leaflets ovate with numerous lobes, acuminate at the apex, upper surface green. Growth vigorous, flowers many. Classic variety.

Rose Form 107

Yin Hong Qiao Dui

Da Hong Duo Jin

Dan Lu Yan

Zhong Sheng Hei

cv. Dan Lu Yan
Rose form. Flowers 13cm × 4cm, at first deep red (44-C), later greyish purple (74-C); petals thin, lustrous, purplish red basal blotches; stamens slightly petaloid; pistils small; floral discs purplish red; stalks slender, stiff, relatively short, flowers upright. Flowering early.

Plant dwarf, partially spreading. Branches slender, pliable. Leaves long, medium-sized; leaflets long elliptic, the margins curved upwards, acute at the apex, pendulous. Growth weak, flowers many, flowers colour gradually fades in sun. Classic variety.

cv. Yin Hong Qiao Dui
Rose form, sometimes chrysanthemum form occurs. Flowers 15cm × 5cm, light red (48-D); petals stiff, suffused with purplish red at the base, regular, conspicuously layered; stamens slightly petaloid, pistils small, floral discs pink. Stalks fairly long, stiff, flowers upright. Flowering midseason.

Plant medium height, partially spreading. Branches stout, stiff. Long leaves, small-sized, stiff, sparse; leaflets long ovate, acuminate at the apex, upper surface yellowish green. Growth vigorous, flowers many, very adaptable, resistant to diseases. Bred by Zhaolou Peony Garden, Heze in 1966.

cv. Da Hong Duo Jin
Rose form . Flowers 18cm × 6cm, deep red (61-A), lustrous; petals multi-whorled, soft, suffused with dark purple at the base; stamens partially petaloid; pistils normal or small; Stalks long, flowers upright. Flowering late.

Plant medium height, erect. Branches relatively slender, stiff. Leaves long, medium-sized, sparse; leaflets long ovate, apex acute, the margins curved upwards. Growth weak, flowers many. Bred by Zhaolou Peony Garden, Heze in 1969.

cv. Zhong Sheng Hei
Rose form. Flowers 13cm × 4cm, pale dark purple (59-C); petals thick, stiff, suffused with dark purple at the base; stamens partially petaloid; pistils normal, sometimes fertile. Stalks stiff, straight; flowers upright. Flowering late midseason.

Plant dwarf, erect. Branches relatively stout, stiff. Leaves long, large-sized, stiff, bi-ternate pinnate; leaflets long ovate, apex acute, the margins curved upwards; young leaflets dark purple, mature leaflets deep green. Growth weak. Classic variety.

cv. Hong Shan Hu

Rose form. Flowers 16cm × 8cm, red (44-C); the outer petals 2-whorled, large, slightly rugose; the inner petals multi-whorled, thin, soft, wrinkled, with numerous teeth, light red at the base, stamens slightly petaloid, pistils small; Stalks long, stiff, flowers upright. Flowering midseason.

Plant dwarf, erect. Branches fairly stiff. Leaves long, medium-sized, biternate pinnate, crowded; leaflets lanceolate, acuminate at the apex, the margins curved upwards, upper surface green. Growth medium; flowers many. Bred by Heze Flower Garden in 1986.

cv. Mei Gui Hong

Rose form, sometimes chrysanthemum form occurs. Flowers 16cm × 5cm, purplish red (53-A), lustrous; petals soft, dark purple basal blotches; stamens slightly petaloid; pistils 7-9; floral discs deep purplish red. Stalks long, stiff, flowers upright. Flowering midseason.

Plant medium height, partially spreading. Branches stout. Leaves orbicular, medium-sized, thick, stiff, fairly sparse; leaflets ovate, shallowly lobed, acute at the apex, the margins curved upwards. Growth vigorous, flowers many, not tolerant of strong sunshine. Bred by Zhaolou Peony Garden, Heze in 1978.

cv. Jiao Yan

Rose form, sometimes lotus form or chrysanthemum form occurs. Flowers 20cm × 5cm, light red (50-C), soft; petals multi-whorled, stiff, flat, layered conspicuously, apex pinkish red, suffused with red at the base; a few stamens petaloid, filaments elongated; pistils nearly normal, floral discs deep purplish red. Stalks stout, stiff, flowers upright. Flowering early midseason.

·Plant medium height, partially spreading. Branches stout. Leaves orbicular, large-sized, thick, sparse; leaflets long elliptic or ovate with a few and shallow lobes, obtuse at the apex. Growth vigorous; flowers many, tolerant of strong sunshine. Bred by Heze Flower Garden in 1973.

cv. Hong Yu

Rose form. Flowers 18cm × 6cm, bright purplish red (60-B); the outer petals 2-whorled, large, stiff, deep purplish red basal blotches; the inner petals sparse, soft, narrowly long, rugose; stamens a few, partially petaloid, pistils developed, petaloid, greenish coloured. Stalks long. Flowers upright or lateral. Flowering late.

Plant medium height, partially spreading. Branches stiff. Leaves orbicular, medium-sized, stiff, thin; leaflets ovate, acute at the apex, the margins curved upwards. Growth vigorous; flowers many; tolerant of low temperature during bud stage. Bred by Heze Flower Garden in 1976.

cv. Hong Yan Bi Rui

Rose form. Flowers 15cm × 5cm, light red (44-D); petals thin, soft, rugose, suffused with red at the base; stamens a few; pistils developed, petaloid, yellowish green coloured. Stalks slender, flowers lateral. Flowering midseason.

Plant medium height, spreading. Branches slender, pliable. Leaves long, small-sized, sparse; leaflets ovate or elliptic, acute at the apex; upper surface yellowish green, the margins slightly curved downwards, suffused with deep purple. Growth weak, leaves falling early, flowers many. Bred by Zhaolou Peony Garden, Heze in 1988.

cv. Shao Nu Qun

Rose form. Flowers 16cm × 5cm, pink somewhat purple (55-B); the outer petals 3-4-whorled, the inner petals wrinkled, suffused deep purple at the base, some stamens petaloid; pistils normal or small; Stalks fairly long, stiff, flowers lateral. Flowering midseason.

Plant tall, erect. Branches slender, stiff, Leaves long, medium-sized, slightly sparse; leaflets long ovate, acuminate at the apex, upper surface green. Growth vigorous; flowers many. Bred by Team 6, Zhaolou, Heze in 1978.

Hong Shan Hu

Mei Gui Hong

Jiao Yan

Hong Yu

Hong Yan Bi Rui

Rose Form 109

Shao Nü Qun

Yan Zi Ying Hui

Hong Xia Zheng Hui

Jin Pao Hong

cv. Yan Zi Ying Hui

Rose form. Flowers 18cm × 7cm, purplish red (61-B); the outer petals 4-5-whorled, large, flat, smooth; the inner petals soft, slightly wrinkled, sparse, suffused with dark purplish red at the base; stamens slightly petaloid; pistils 9-11; floral discs pinkish white. Stalks long, stiff, pale purple, flowers upright or lateral. Flowering late.

Plant tall, partially spreading. Branches stout. Leaves long, medium-sized, relatively crowded; leaflets ovate lanceolate, acuminate at the apex, the margins curved upwards, upper surface deep green suffused with pale purple. Growth vigorous. Bred by Heze Flower Garden in 1994.

cv. Hong Xia Zheng Hui

Rose form. Flowers 17cm × 6cm, purplish red faintly bluish (53-B); petals multi-whorled, soft, crowded, dark purple basal blotches; stamens slightly petaloid, pistils small or petaloid. Stalks long, flowers lateral. Flowering midseason.

Plant high, partially spreading. Branches slender. Leaves long, medium-sized, sparse; leaflets ovate or long ovate with numerous lobes, acute at the apex. Growth vigorous; flowers many. Bred by Zhaolou Peony Garden, Heze in 1963.

cv. Jin Pao Hong

Rose form, sometimes chrysanthemum form occurs. Flower buds oblate, acute at the top; flowering purplish red (64-C); petals multi-whorled, arranged closely, gradually smaller from the outer to the inner petals, lighter colour at the apex, dark purple basal blotches; stamens normal or slightly petaloid; pistils small. Stalk short, stiff, flowers upright. Flowering midseason.

Plant tall, partially spreading. Branches stout. Leaves orbicular, large-sized, relatively crowded; leaflets broadly ovate or elliptic with numerous lobes, apex obtuse, suffused with deep purple on the margins. Growth vigorous, flowers many. Classic variety.

cv. Tao Li Zheng Yan

Rose form, sometimes chrysanthemum form occurs. Flowers 16cm × 5cm, red (52-D), pink at the apex; petals stiff, dark purple basal blotches, arranged regularly; stamens slightly petaloid; pistils numerous, floral discs deep purple. Stalks fairly short, stiff, flowers upright. Flowering midseason.

Plant medium height, partially spreading. Branches fairly stout. Leaves long, medium-sized, stiff, crowded; leaflets long ovate, the margins slightly curved upwards, acute at the apex. Growth medium; flowers many, tolerant of strong sunshine. Bred by Zhaolou Peony Garden, Heze in 1982.

cv. Mu Chun Hong

Rose form, sometimes chrysanthemum form occurs. Flowers 20cm × 6cm, pale purplish red (63-B); petals multi-whorled, outer ones 2-3-whorled, stiff, flat, the inner ones gradually smaller, slightly twisted upwards, suffused with deep purple at the base; stamens slightly petaloid, pistils 5-7, occasionally petaloid, floral discs purplish red. Stalks stiff, fairly long, flowers upright or lateral. Flowering late.

Plant medium height, partially spreading. Branches fairly stiff. Leaves orbicular, medium-sized, slightly crowded, soft; leaflets ovate, acute at the apex, upper surface smooth, yellowish green. Growth vigorous, flowers many. Bred by Heze Flower Garden in 1988.

Rose Form 111

Tao Li Zheng Yan

Mu Chun Hong

cv. Hong Bao Shi

Rose form. Flowers 16cm × 5cm, red (44-C), smooth; petals multi-whorled, soft, crowded, suffused with purplish red at the base; stamens slightly petaloid; pistils small. Stalks long, stiff, flowers upright. Flowering midseason.

Plant medium height, partially spreading. Branches stout. Leaves long, medium-sized, stiff, crowded; leaflets ovate or long ovate, acuminate at the apex, the upper surface relatively rough, green. Growth vigorous; flowers many. Bred by Zhaolou Peony Garden, Heze in 1973.

Hong Bao Shi

cv. Du Juan Hong

Rose form. Flowers 17cm × 5cm, purplish red (61-B), bright-colour; petals multi-whorled, broad and large, wrinkled at the apex, suffused with deep purplish red at the base; some stamens petaloid, narrowly long, rugose; pistils 5-7; floral discs purplish red. Stalks long, stiff, flowers upright. Flowering late.

Plant tall, partially spreading. Branches fairly slender, stiff, slightly curved. Leaves orbicular, medium-sized, sparse; leaflets ovate, acute at the apex, pendulous, the mar-gins slightly curved upwards, suffused with purple. Growth medium, flowers many, produced in profusion, tolerant of strong sunshine, flowering period late. Bred by Zhaolou Peony Garden, Heze in 1984.

Du Juan Hong

Rose Form

Mei Gui Zi

cv. Mei Gui Zi
Rose form. Flowers 16cm × 6cm, purple (71-D); petals multi-whorled, stiff, light coloured at the apex, dark purple basal blotches; stamens partially petaloid, small. Stalks stiff, flowers upright. Flowering late midseason.

Plant tall, erect. Branches relatively stout. Leaves long, medium-sized, sparse; leaflets long ovate or lanceolate, apex acuminate, the margins curved upwards, upper surface suffused with greenish purple. Growth vigorous, flowers many; not tolerant of leaf spot, leaves falling early. Bred by Zhaolou Peony Garden, Heze in 1965.

cv. Wu Jin Yao Hui
Rose form, sometimes chrysanthemum form occurs. Flowers 16cm × 5cm, dark purplish red (187-D), lustrous; petals multi-whorled, outer petals 4-whorled, flat, the inner petals soft, rugose, dark purple basal blotches; stamens partially petaloid, pistils small. Stalks slender, stiff, long, flowers upright or lateral. Flowering midseason.

Plant medium height, partially spreading. Branches slender. Leaves long, medium-sized; leaflets long ovate, with a few lobes, the margins curved upwards, suffused with pale purple; upper surface deep green. Growth medium; flowers many; tolerant of adverse conditions. Bred by Zhaolou Peony Garden, Heze in 1980.

Wu Jin Yao Hui

Hundred Proliferate-Flower Form

Two or more individual flowers of Hundred proliferate form overlapping.

Cao Zhou Hong

cv. Cao Zhou Hong

Hundred proliferate form. Flowers 17 cm × 9cm, red (50-B). Petals 6-7 whorled, stiff, flat, regular, suffused with light red, stamens a few, pistils developed, petaloid, greenish coloured in the lower flower. Petals relatively few, wrinkled; stamens small; pistils small. Stalks stiff, flowers upright. Flowering early.

Plant medium height, partially spreading. Branches stout. Leaves orbicular, medium-sized, crowded, thick, stiff; leaflets broadly-ovate, acute at the apex. Growth vigorous; flowers many; tolerant of adverse conditions. Bred by Heze Flower Garden in 1982.

cv. Luo Du Chun Yan

Hundred proliferate form. Flowers 15 cm × 9cm, red (52-C). The outer petals 3-4-whorled, broad and large, slightly wrinkled, the inner petals gradually smaller, arranged regularly; pistils developed, mixed red and green petals; stamens partially petaloid in the lower flower. Petals irregular, wrinkled, erect, stamens partially petaloid, pistils small in the upper flower. Stalks straight, flowers upright. Flowering late midseason.

Plant medium height, erect. Branches stout. Leaves orbicular, medium-sized, crowded; leaflets ovate, obtuse at the apex. Growth vigorous, flowering medium. Bred by Wangcheng Park, Luoyang in 1969.

cv. Yan Zhi Dian Cui

Hundred proliferate form. Flowers 19 cm × 12cm, red (52-A). Petals narrow, fairly long, the majority of them wrinkled, arranged irregularly, the majority of stamens petaloid; pistils developed, mixed red and green coloured petals in the lower flower. Petals fairly large, 4-5-whorled; stamens partially petaloid; pistils normal; floral discs purplish red in the upper flower. Stalks fairly stiff, flowers lateral. Flowering midseason.

Plant medium height, erect. Branches fairly stout. Leaves long, large-sized, crowded; leaflets long elliptic, acuminate at the apex. Growth very vigorous, flowers normal. Bred by Wangcheng Park, Luoyang in 1969.

Hundred Proliferate-Flower Form 115

Yan Zhi Dian Cui

Luo Du Chun Yan

cv. Hong Hua Lu Shuang

Hundred proliferate form. Flowers 18 cm × 7cm, red (47-C). Petals multi-whorled, gradually shorter from the outer to the inner whorls, purplish red basal blotches; stamens a few; pistils developed, petaloid, reddish coloured in the lower flower. Petals large, erect, a few; stamens small, pistils small or slightly petaloid in the upper flower. Stalks stiff, flowers upright. Flowering midseason.

Plant tall, erect. Branches stiff. Leaves long, small-sized, thick, stiff, sparse; leaflets long ovate, acute at the apex. Growth vigorous, flowers many; tolerant of adverse conditions. Bred by Heze Flower Garden in 1976.

Hong Hua Lu Shuang

cv. Jun Yan Hong

Hundred proliferate form. Flowers 18 cm × 7cm, pinkish purple bluish (73-B). Petals 3-4-whorled, broad and large, stiff, the margins wavily toothed, suffused with purplish red at the base, stamens a few, occasionally petaloid; pistils developed, petaloid, reddish green coloured in the lower flower. Petals relatively few, rugose, toothed at the apex, long, straight, stamens small, pistils small in the upper flower. Stalks stout, long, straight; flowers upright. Flowering midseason.

Plant tall, erect. Branches relatively stout. Leaves long, medium-size, thick; leaflets long ovate, acuminate at the apex. Growth vigorous, flowers many; tolerant of leaf spot, leaves falling late. Bred by Zhaolou Peony Garden, Heze in 1969.

Jun Yan Hong

Hundred Proliferate-Flower Form 117

Ying Hong

Lan Bao Shi

cv. Ying Hong

Hundred proliferate form. Flowers 18 cm × 6cm, red bluish (57-C). Petals 5-7-whorled, with translucent streaks, dark purple basal blotches, stamens many; pistils developed, petaloid, greenish coloured in the lower flower. Petals relatively few, rugose; stamens small; pistils small in the upper flower. Stalks stiff, flowers upright. Flowering midseason.

Plant dwarf, partially spreading. Branches stout. Leaves long, medium-sized, thick, relatively crowded; leaflets ovate or broadly ovate-lanceolate, the margins curved upwards, acute at the apex, slightly pendulous, upper surface deep green suffused with pale purple. Growth medium, flowers many. Bred by Zhaolou Peony Garden, Heze in 1968.

cv. Lan Bao Shi

Hundred proliferate form, sometimes chrysanthemum form occurs. Flowers 16cm × 6cm, pink bluish (65-B). Petals 5-7-whorled, stiff, outer petals 2-whorled, flat, dark purple basal blotches, the inner petals slightly rugose; pistils developed, petaloid, greenish coloured in the lower flower. Petals relatively few, stiff, apex light coloured in the upper flower. Stalks long, stiff, flowers upright. Flowering midseason.

Plant medium height, erect. Branches stiff. Leaves long, medium-sized, crowded, stiff; leaflets ovate or long ovate, acuminate at the apex, the margins curved upwards. Growth medium, flowers many; tolerant of strong sunshine. Bred by Team 9, Zhaolou, Heze in 1975.

Xin Jiao Hong

Gui Fei Cha Cui

Ying Ri Hong

Fei Yan Hong Zhuang

Lü Die Wu Fen Lou

Hong Tu

cv. Xin Jiao Hong
Hundred proliferate form. Flowers 16 cm × 5cm, red (43-C), soft. Petals 5-6-whorled, soft, sparse, curved inward, purplish red basal blotches, stamens a few, pistils developed, petaloid, yellowish green coloured in the lower flower. Petals relatively few, twisted; stamens and pistils small or pistils petaloid in the upper flower. Stalks pliable, flowers lateral. Flowering midseason.

Plant tall, spreading. Branches stout, pliable. Leaves orbicular, medium-sized, thick, sparse; leaflets ovate, apex acute. Growth vigorous, flowers many. Bred by Zhaolou Peony Garden, Heze. in 1972.

cv. Hong Tu
Hundred proliferate form. Flowers 17 cm × 5cm, light red (40-D). Petals multi-whorled, conspicuously smaller from the outer to the inner petals, thin, soft, rolled and wrinkled, stamens a few, pistils small in the lower flower. Petals relatively few, fewer stamens and pistils in the upper flower. Stalks short, flowers slightly hidden among the leaves. Flowering midseason.

Plant dwarf, spreading. Branches fairly stout. Leaves long, medium-size, crowded; leaflets narrowly long, ovate lanceolate, acuminate at the apex, lateral lobes pendulous. Growth vigorous; flowers many, not tolerant of strong sunshine. Bred by Zhaolou Peony Garden, Heze in1969.

cv. Gui Fei Cha Cui
Hundred proliferate form. Flowers 15 cm × 7cm, pinkish red (52-D). Petals 5-6-whorled, arranged regularly, base suffused with red; stamens a few; pistils developed, petaloid, yellowish green coloured in the lower flower. Petals relatively few, rugose, irregularly toothed at the apex, raised, stamens and pistils small or petaloid in the upper flower. Stalks long, stiff, flowers upright. Flowering midseason.

Plant tall, erect. Branches stout. Leaves orbicular, medium-sized, thick, fairly sparse; leaflets ovate, acute at the apex. Growth vigorous; flowers many. Bred by Team 9, Zhaolou, Heze in 1970.

cv. Lu Die Wu Fen Lou
Hundred proliferate form. Flowers 15cm × 8cm, pinkish red (52-D). Petals 5-whorled, stiff, often toothed at the apex, stamens slightly petaloid; pistils developed, petaloid, greenish coloured in the lower flower. Petals relatively few, wrinkled, stamens and pistils small in the upper flower. Stalks fairly stiff, flowers upright or lateral. Flowering midseason.

Plant medium height, partially spreading. Branches stiff. Leaves orbicular, medium-sized, crowded; leaflets ovate, acute at the apex. Growth medium; flowers many, tolerant of strong sunshine. Bred by Zhaolou Peony Garden, Heze in 1988.

cv. Ying Ri Hong
Hundred proliferate form. Flowers 17 cm × 6cm, red (52-C), later light pink at the apex. The outer petals 4-whorled, entire and regular, flat, stiff, arranged regularly, dark purple basal blotches, stamens partially petaloid; pistils developed, petaloid, greenish coloured in the lower flower. Petals relatively few, large, stamens a few, pistils small in the upper flower. Stalks long, straight, flowers upright. Flowering early.

Plant medium height, erect. Branches fairly stout. Leaves long, medium-sized, stiff; leaflets long ovate, acuminate at the apex. Growth vigorous, flowers many. Bred by Team 9, Zhaolou, Heze in 1970.

cv. Fei Yan Hong Zhuang
Hundred proliferate form. Flowers 16 cm × 6cm, red (52-C), pink at top. Petals multi-whorled, slightly soft, dark purple basal blotches; stamens small; pistils developed, petaloid, greenish coloured in the lower flower. Petals relatively few, slightly large; stamens small; pistils small in the upper flower. Stalks slender and long, flowers lateral. Flowering midseason.

Plant medium height, spreading. Branches slender, pliable. Leaves long, small-sized, sparse; leaflets elliptic, terminal leaflets irregularly toothed. Growth vigorous, flowers many. Bred by Zhaolou Peony Garden, Heze in 1964.

Hong Lou Chun Hui

Tao Hong Fei Cui

Ling Hua Zhan Lu

Huo Lian Bi Yu

Shi Ba Hao

Zhi Hong

cv. Hong Lou Chun Hui

Hundred proliferate form. Flowers 17 cm × 10cm, magenta (52-A), lustrous. The outer petals 7-8-whorled, broad and large, entire and regular, gradually smaller from the outer to the inner petals, purple at the base; stamens normal or occasionally petaloid, pistils developed, mixed red and green coloured petals in the lower flower. Petals 3-whorled, broad and large, erect, wrinkled, pink at the apex, stamens normal, pistils small. Stalks fairly pliable, flowers lateral. Flowering midseason.

Plant dwarf, partially spreading. Leaves orbicular, medium-sized; leaflets broadly ovate, obtuse at the apex, pendulous. Growth medium, flowers many. Bred by Wangcheng Park, Luoyang in 1969.

cv. Huo Lian Bi Yu

Hundred proliferate form. Flower buds conical, often opening; flowers 16cm × 5 cm, magenta (43-B), lustrous. Petals 6-whorled, slightly large, lighter colour at the apex; stamens occasionally petaloid; pistils developed, petaloid, greenish coloured in the lower flower. Petals relatively few, twisted and wrinkled, stamens small; pistils small in the upper flower. Stalks short, flowers upright or lateral. Flowering late.

Plant dwarf, partially spreading. Branches stout. Leaves long, medium-sized, crowded; leaflets ovate, apex acute, pendulous. Growth vigorous, flowers relatively few. Bred by Zhaolou Peony Garden, Heze in 1969.

cv. Tao Hong Fei Cui

Hundred proliferate form. Flowers 20 cm × 10cm, deep pinkish red (55-B). The outer petals 3-4-whorled, broad and large, the inner petals wrinkled; stamens slightly petaloid; pistils developed, petaloid, greenish coloured in the lower flower. Petals relatively few, rugose; stamens small; pistils small in the upper flower. Stalks stout, slightly pliable, flowers lateral. Flowering midseason.

Plant tall, partially spreading. Branches stout. Leaves long, large-sized, slightly thick; leaflets long ovate, apex acuminate. Growth vigorous, flowers many. Bred by Zhaolou Peony Garden, Heze in 1978.

cv. Shi Ba Hao

Hundred proliferate form, sometimes chrysanthemum form occurs. Flowers 20cm × 8cm, red (58-C). Petals 6-7-whorled, stiff, deep purplish red basal blotches; stamens slightly petaloid; pistils small or petaloid in the lower flower. Petals relatively few, erect, fairly large; stamens small; pistils small in the upper flower. Stalks stout, long, flowers upright. Flowering midseason.

Plant tall, erect. Branches stout. Leaves orbicular, medium-sized, thick, stiff; leaflets nearly orbicular, apex obtuse. Growth vigorous, flowers many. Classic variety.

cv. Ling Hua Zhan Lu

Hundred proliferate form. Flowers 20 cm × 6cm, pinkish purple (73-B), later pinkish white at the apex. Petals 8-whorled, thin, soft; stamens slightly petaloid; pistils small or developed, petaloid, light greenish coloured in the lower flower. Petals multi-whorled, wrinkled; stamens smaller or absent; pistils small or absent in the upper flower. Stalks fairly long, flowers lateral. Flowering late.

Plant medium height, spreading. Branches stout. Leaves long, medium-sized, crowded, horizontal, thin, soft, pendulous; leaflets ovate or broadly lanceolate, acuminate at the apex. Growth vigorous; flowers many. Bred by Zhaolou Peony Garden, Heze in 1969.

cv. Zhi Hong

Hundred proliferate form. Flowers 16 cm × 6cm, magenta (44-C), smooth. The outer petals 2-3-whorled, flat, the inner petals soft, wrinkled, arranged closely, purplish red basal blotches; stamens relatively few; pistils small in the lower flower. Petals relatively few, soft, wrinkled; pistils small in the upper

Lan Fu Rong

flower. Stalks short, flowers lateral. Flowering midseason.

Plant medium height, spreading. Branches stout. Leaves long, medium-sized, fairly crowded; leaflets long ovate, acuminate at the apex. Growth vigorous, flowers many, not tolerant of strong sunshine. Classic variety.

cv. Lan Fu Rong

Hundred proliferate form. Flowers 20 cm × 8cm, pink bluish (73-C). Petals multi-whorled, relatively large, suffused with purple at the base; stamens occasionally petaloid; pistils small in the lower flower. Petals relatively few, narrowly long, wrinkled; stamens small; pistils small in the upper flower. Stalks long, straight, flowers upright. Flowering late midseason.

Plant tall, erect. Branches stout. Leaves long, medium-sized, thick; leaflets ovate, apex acuminate. Growth vigorous, flowers many; resistant to leaf spot. Bred by Zhaolou Peony Garden, Heze in 1969.

Lan Fu Rong

Hundred Proliferate-Flower Form

Xi Shi Huan Sha

Tao Hua Jiao Yan

Zi Yun

cv. Xi Shi Huan Sha
Hundred proliferate form. Flowers 19 cm × 9cm, pinkish red bluish (73-C); petals 8-whorled, large, flat, purplish red basal blotches; stamens slightly petaloid; pistils developed, petaloid, greenish coloured in the lower flower. Petals relatively few, large, rugose; stamens small; pistils small in the upper flower. Stalks long, stiff, flowers upright. Flowering midseason.
Plant tall, erect. Branches stout; branch-lets and internodes long. Leaves long, small-sized, crowded; leaflets broadly lanceolate or ovate, acute at the apex. Growth vigorous, flowers many. Bred by Zhaolou Peony Garden, Heze in 1988.

cv. Tao Hua Jiao Yan
Hundred proliferate form. Flowers 19 cm × 7cm, light red (48-C). Petals 3-4-whorled, thin, flat, suffused with red at the base; stamens relatively few; pistils developed, petaloid, greenish coloured in the lower flower. Petals relatively few, rugose, soft, deeper colour; stamens relatively few, small; pistils small or slightly petaloid in the upper flower. Stalks stiff, flowers upright or lateral. Flowering midseason.
Plant tall, partially spreading. Branches stout, stiff. Leaves orbicular, medium-sized, soft, thick, slightly sparse; leaflets ovate or long ovate, acute at the apex, the margins slightly wavy. Growth vigorous. Bred by Heze Flower Garden in 1969.

cv. Zi Yun
Hundred proliferate form, sometimes chrysanthemum form occurs. Flowers 18cm × 7cm, purplish red (61-C); petals 6-8-whorled, stiff, regular, dark purple basal blotches, stamens normal, slightly petaloid; occasionally pistils developed, petaloid, pinkish white coloured in the lower flower. Petals sparse; stamens a few; pistils small or developed, petaloid, pinkish yellow coloured in the upper flower. Stalks stout, flowers upright. Flowering midseason.
Plant medium height, partially spreading. Branches fairly stout. Leaves orbicular, medium-sized, thick, stiff; leaflets ovate, acute at the apex. Growth vigorous, flowers many. Bred by Zhaolou Peony Garden, Heze in 1983.

Qing Long Zhen Bao

Hundred proliferate form. Flowers 18 cm × 8cm, purplish red (61-C). Petals 4-5-whorled, large, flat, suffused with dark purple at the base; stamens small; pistils developed; petaloid, yellowish green coloured in the lower flower. Petals relatively few, rugose; stamens small; pistils small in the upper flower. Stalks slender, fairly pliable, pale purple, flowers lateral. Flowering late midseason.

Plant tall, spreading. Branches slender, slightly pliable. Leaves long, medium-sized, soft, sparse; leaflets elliptic or ovate, apex acuminate. Growth vigorous, flowers many. Bred by Heze Flower Garden in 1977.

cv. Qing Shan Wo Yun

Hundred proliferate form. Flowers 14 cm × 10cm, pinkish purple, faintly bluish (68-B). The outer petals large, arranged irregularly, the inner petals very small, wrinkled; filaments partially petaloid, anthers deformed and remained; pistils developed, mixed purple and green coloured petals in the lower flower. Petals narrow, slightly rugose; stamens normal or petaloid; pistils normal or petaloid; floral discs white, stigmas red in the upper flower. Stalks pliable, flowers lateral. Flowering midseason.

Plant tall, partially spreading. Branches stout. Leaves orbicular, small-sized; leaflets elliptic, obtuse at the apex. Growth medium, flowers many.

cv. Sheng Ge Jin

Hundred proliferate form. Flowers 18 cm × 6cm, pinkish purple (73-C). Petals 6-8-whorled, large, flat, suffused with purplish red at the base; stamens relatively few, petaloid, short, small, wrinkled; pistils developed, petaloid, greenish coloured in the lower flower. Petals multi-whorled, broad and large, incurved toward the flower centre, stamens small or absent; pistils small in the upper flower. Stalks stout, straight. Flowering midseason.

Plant medium height, partially spreading. Branches stout. Leaves long, medium-sized; leaflets long ovate, apex acuminate. Growth vigorous, flowers many. Bred by Zhaolou Peony Garden, Heze in 1967.

Qing Long Zhen Bao

cv. Yan Zhu Jian Cai

Hundred proliferate form. Flowers 17 cm × 10cm, pale purplish red (57-D). Petals multi-whorled, arranged irregularly, wrinkled, shallowly toothed at the apex, suffused with red at the base; stamens a few; pistils slightly petaloid in the lower flower. Petals relatively few, wrinkled, erect, lighter colour at the apex; stamens small; pistils small in the upper flower. Stalks stiff, flowers lateral. Flowering midseason.

Plant medium height, partially spreading. Branches fairly stiff. Leaves long, medium-sized; leaflets long ovate, acuminate at the apex. Growth vigorous, flowers many. Bred by Zhaolou Peony Garden, Heze in 1963.

cv. Zi Ge

Hundred proliferate form. Flowers 18 cm × 9cm, purplish red (64-C). The outer petals 4-5-whorled, broad and large, ovate, purple at the base; stamens partially petaloid; pistils developed, petaloid, greenish coloured in the lower flower. Stamens normal; pistils normal in the upper flower. Stalks stout, straight. Flowering midseason.

Plant medium height, partially spreading. Branches stout. Leaves long, large-sized; leaflets long elliptic or lanceolate, acute at the apex. Growth vigorous, flowers many. Bred by Luoyang Peony Garden, Luoyang in 1995.

cv. Lu He Hong

Hundred proliferate form. Flowers 18 cm × 7cm, red (63-B). Petals 7-8-whorled, large, suffused with purplish red at the base; stamens partially petaloid; pistils small in the lower flower. Petals multi-whorled, large, rugose; stamens a few, pistils small in the upper flower. Stalks straight, slightly short, flowers upright. Flowering midseason.

Plant medium height, partially spreading. Branches stout. Leaves long, medium-sized, thick, crowded; leaflets long elliptic, acute at the apex. Growth vigorous, flowers many, leaf falling late. Bred by Zhaolou Peony Garden, Heze in 1968.

cv. Ni Hong Huan Cai

Hundred proliferate form. Flowers 15 cm × 7cm, magenta (43-B). Petals multi-whorled, arranged regularly, dark purple basal blotches, stamens a few, pistils developed, petaloid, light greenish coloured in the lower flower. Petals slightly large, rugose; stamens small; pistils small in the upper flower. Stalks slightly long, flowers upright or lateral. Flowering midseason.

Plant tall, partially spreading. Branches stout. Leaves orbicular, medium-sized, crowded, thick; leaflets ovate, acute at the apex. Growth vigorous, flowers many. Bred by Team 9, Zhaolou, Heze in 1972.

Qing Shan Wo Yun

cv. Tao Hua Xi Jin
Hundred proliferate form. Flowers 17 cm × 8cm, deep pink. petals 7-8-whorled, broad and large, slightly wrinkled, the margins toothed; stamens normal or smaller; pistils developed, petaloid, normal, greenish on the apical margins in the lower flower. Petals large, multi-whorled; stamens a few; pistils small in the upper flower. Stalks stout, straight, flowers upright. Flowering midseason.
Plant medium height, partially spreading. Branches stout. Leaves orbicular, small-sized, crowded; leaflets broadly obovate, deep lobed or completely cut, acuminate at the apex. Growth vigorous, flowers many. Bred by Zhaolou Peony Garden, Heze in 1987.

cv. Tian Ran Fu Gui
Hundred proliferate form. Flowers 19 cm × 11cm, deep purplish red (61-B), lustrous. The outer petals 3-whorled, broad and large, the margins slightly wrinkled, deep purplish red at the base; stamens partially petaloid; pistils developed, petaloid, greenish coloured in the lower flower. Petals fairly small, wrinkled, stamens normal; pistils normal; floral discs deep purple in the upper flower. Stalks fairly straight, flowers upright. Flowering midseason.
Plant tall, erect. Branches stout. Leaves long, medium-sized; leaflets long ovate, acuminate at the apex. Growth vigorous, flowers many. Bred by Guose Peony Garden, Luoyang in 1992.

cv. Mo Jian Rong
Hundred proliferate form. Flowers 15 cm × 6cm, dark purple (187-D); petals 6-8-whorled, conpiscuously different in size from outer to inner petals, suffused with black at the base; stamens slightly petaloid; pistils developed, petaloid, yellowish green coloured in the lower flower. Petals multi-whorled, large; stamens small; pistils small in the upper flower. Stalks fairly long, stiff, flowers lateral. Flowering midseason.
Plant medium height, partially spreading. Branches stout. Leaves long, small-sized; leaflets long ovate or lanceolate, terminal leaflets completely cut or the one side completely cut, acute at the apex. Growth vigorous, flowers many. Bred by Zhaolou Peony Garden, Heze in 1983.

cv. Wu Long Peng Sheng
Hundred proliferate form. Flowers 16 cm × 6cm, purplish red (53-B), lustrous. Petals multi-whorled, the outer ones 2-whorled, large, stiff, gradually smaller from outer to inner petals; stamens a few; pistils developed, mixed red and green coloured petals in the lower flower. Petals relatively few, rugose; stamens smaller or petaloid; pistils small or slightly petaloid in the upper flower. Stalks fairly long, flowers upright or lateral. Flowering midseason.
Plant tall, partially spreading. Branches stout, stiff. Leaves long, medium-sized, stiff, sparse; leaflets ovate or long ovate, acute at the apex. Growth vigorous; flowers many. Classic variety.

cv. Hong Hui
Hundred proliferate form. Flowers 14 cm × 7cm, pale purple (57-D). Petals 3-4-whorled, irregularly toothed at the apex, dark purple basal blotches; stamens occasionally petaloid; pistils slightly small in the lower flower. Petals relatively few, with numerous and rugose teeth; stamens small; pistils small in the upper flower. Stalks slender, stiff, flowers upright. Flowering midseason.
Plant medium height, erect. Branches fairly slender; branchlets long. Leaves long, small-sized, crowded, stiff; leaflets long ovate, acuminate at the apex. Growth medium; flowers many. Bred by Zhaolou Peony Garden, Heze in 1967.

cv. Tao Hua Zheng Yan
Hundred proliferate form. Flowers 14 cm × 10cm, pale purplish red (67-B). The outer petals 5-6-whorled, large, flat, crowded; stamens partially petaloid; pistils developed, petaloid, coloured in the lower flower. Petals large, erect, rugose; stamens occasionally petaloid; pistils normal in the upper flower. Pedicels erect, flowers upright. Flowering midseason.
Plant dwarf, erect. Branches slender, stiff. Leaves long, medium-sized, sparse; leaflets long ovate, acute at the apex. Growth medium, flowers many. Bred by Wangcheng Park, Luoyang in 1969.

cv. Dong Fang Jin
Hundred proliferate form. Flowers 17 cm × 6cm, pale purplish red (67-B). Petals multi-whorled, the margins toothed, stamens partially petaloid, pistils small in the lower flower. Petals sparse, erect; stamens a few; pistils small in the upper flower. Stalks stout, long, flowers upright. Flowering midseason.
Plant medium height, partially spreading. Branches fairly stout, stiff. Long leaves, medium-sized; leaflets ovate, acute at the apex. Growth vigorous, flowers relatively few. Bred by Zhaolou Peony Garden, Heze in 1966.

124 Chinese Tree Peony

Sheng Ge Jin

Lu He Hong

Yan Zhu Jian Cai

Zi Ge

Ni Hong Huan Cai

Tao Hua Xi Jin

Hundred Proliferate-Flower Form 125

Tian Ran Fu Gui

Hong Hui

Mo Jian Rong

Tao Hua Zheng Yan

Dong Fang Jin

Wu Long Peng Sheng

cv. Chun Gui Hua Wu

Hundred proliferate form. Flowers 17 cm × 11cm, pale purplish red (68-A). The outer petals large, slightly curved downwards; anthers are always at the apex of stamens; stigmas developed, petaloid, small, greenish coloured in the lower flower. Petals fairly large, wrinkled, erect; stamens normal; pistils normal, stigma red; floral discs purplish red in the upper flower. Stalks short, straight. Flowering midseason.

Plant dwarf, partially spreading. Branches stout. Leaves long, medium-sized, crowded; leaflets long elliptic, apex acute. Growth vigorous, flowers many. Bred by Guose Peony Garden, Luoyang in 1992.

Chun Gui Hua Wu

Chun Gui Hua Wu

Chao Yi

cv. Chao Yi

Hundred proliferate form, sometimes rose form occurs. Flowers 19cm × 10cm, purplish red (61-C); petals 6-8-whorled, large, stamens a few, staminode, small, long; pistils developed, petaloid, greenish coloured in the lower flower. Petals multi-whorled, slightly large, curved; stamens a few, smaller in the upper flower. Stalks fairly long, slightly pliable, flowers lateral. Flowering late.

Plant tall, partially spreading. Branches stout. Leaves long, medium-sized, slightly soft; leaflets long ovate or long elliptic; terminal leaflets deeply trilobed, the terminal lobes deeply trilobed. Growth vigorous, flowers many. Bred by Zhaolou Peony Garden, Heze in 1982.

Chao Yi

Hundred Proliferate-Flower Form 127

Wan Hua Sheng

Yi Pin Zhu Yi

Wu Zhou Hong

Hong Xia Ying Ri

cv. Wan Hua Sheng

Hundred proliferate form. Flowers 20 cm × 9cm, red (58-C), bright. Petals 8-whorled, nearly entire and regular, dark purple basal blotches; stamens a few, occasionally petaloid; pistils small or petaloid in the lower flower. Petals relatively few, wrinkled; stamens small; pistils small in the upper flower. Stalks stout, stiff, long; flowers upright. Flowering late.

Plant tall, erect. Branches stout. Leaves orbicular, large-sized, thick, sparse; leaflets broadly ovate, acute at the apex. Growth vigorous, flowers relatively few. Classic variety.

Jiu Du Zi

cv. Yi Pin Zhu Yi

Hundred proliferate form. Flowers 15 cm × 5cm, pinkish red at the apex (43-B). Petals 6-7-whorled, arranged regularly and crowded, purplish red basal blotches; stamens numerous; stigmas enlarged in the lower flower. Petals very small, a few; stamens small; pistils small in the upper flower. Stalks short, flowers lateral. Flowering midseason.

Plant dwarf, spreading. Branches slender, pliable. Leaves orbicular, small-sized, crowded; leaflets ovate, acuminate at the apex. Growth medium; flowers many. Classic variety.

cv. Wu Zhou Hong

Hundred proliferate form. Flowers 17 cm × 9cm, purplish red (63-A). Petals 2-3-whorled, large, flat, the inner petals small, wrinkled, suffused with purple at the base; stamens slightly petaloid, pistils small or petaloid in the lower flower. Petals large, a few, rugose; stamens small; pistils small in the upper flower. Stalks long, stiff, flowers upright. Flowering midseason.

Plant tall, erect. Branches stiff. Leaves long, medium-sized, stiff, sparse; leaflets ovate, acute at the apex. Growth vigorous; flowers many. Bred by Zhaolou Peony Garden, Heze in 1966.

cv. Hong Xia Ying Ri

Hundred proliferate form, sometimes chrysanthemum form occurs. Flowers 16cm × 6cm, purplish red (61-C), lustrous. Petals multi-whorled, stiff, arranged regularly; stamens slightly petaloid; pistils small or petaloid in the lower flower. Petals relatively few, toothed at the apex; stamens small; pistils small in the upper flower. Stalks fairly long, flowers upright. Flowering midseason.

Plant dwarf, partially spreading. Branches stout. Leaves long, medium-sized, crowded; leaflets long elliptic, apex acute. Growth fairly vigorous; flowers relatively few. Bred by Zhaolou Peony Garden, Heze in 1976.

cv. Jiu Du Zi

Hundred proliferate form. Flowers 17 cm × 11cm, pale purple (67-B), slightly lustrous. The outer petals 6-7-whorled, flat, entire and regular; stamens normal; pistils developed, mixed purple and green coloured petals in the lower flower. Petals fairly large, erect, the margins lobed; stamens normal; pistils normal, stigmas red; floral discs greenish white. Stalks straight, flowers slightly lateral. Flowering late.

Plant tall, erect. Branches stout. Leaves long, medium-sized; leaflets nearly lanceolate, apex acuminate, curved downwards. Growth medium, flowers many. Bred by Wangcheng Park, Luoyang in 1969.

Chong Lou Dian Cui

cv. Chong Lou Dian Cui

Hundred proliferate form. Flowers 16 cm × 7cm, red (57-D). Petals multi-whorled, stiff, suffused with purplish red at the base; stamens occasionally petaloid; pistils developed, petaloid, greenish coloured in the lower flower. Petals relatively few, toothed at the apex; stamens small; pistils small in the upper flower. Stalks stiff, flowers upright. Flowering late.

Plant medium height, partially spreading. Branches stout. Leaves long, medium-sized; leaflets elliptic or ovate elliptic, acute at the apex. Growth vigorous, flowers many. Bred by Zhaolou Peony Garden, Heze in 1968.

cv. Fu Gui Man Tang

Hundred proliferate form, sometimes chrysanthemum form occurs. Flowers 20cm × 6cm, pinkish red faintly bluish (73-A). Petals 7-8-whorled, stiff, flat, arranged regularly, suffused with purple at the base; stamens a few, occasionally petaloid; pistils small in the lower flower. Petals relatively few, narrowly long, twisted; stamens small; pistils small in the upper flower. Stalks fairly stiff, flowers lateral. Flowering midseason.

Plant medium height, partially spreading. Branches stout. Leaves long, medium-sized, stiff; leaflets long ovate, the terminal ones long, completely cut, acuminate at the apex. Growth vigorous, flowers many. Bred by Zhaolou Peony Garden, Heze in 1983.

cv. Li Yuan Chun

Hundred proliferate form. Flowers 16 cm × 7cm, pinkish red (55-A). The outer petals 2-3-whorled, large, slightly spreading, the inner petals wrinkled, sparse; stamens a few; pistils developed, petaloid, greenish coloured in the lower flower. Petals relatively few, thin, rugose; stamens small; pistils small in the upper flower. Stalks slender, stiff, flowers upright. Flowering midseason.

Plant medium height, erect. Branches slender, stiff. Leaves long, small-sized, stiff, sparse; leaflets elliptic, acuminate at the apex. Growth medium, flowers many. Bred by Liji Peony Garden, Heze in 1984.

Hundred Proliferate-Flower Form 129

Fu Gui Man Tang

Li Yuan Chun

Peng Sheng Zi

Peng Sheng Zi

cv. Peng Sheng Zi

Hundred proliferate form. Flowers 15 cm × 5cm, purplish red (53-B), lustrous. Petals multi-whorled, slightly large, stiff, suffused with dark purple at the base; stamens a few; pistils developed, petaloid, greenish coloured in the lower flower. Petals relatively few, wrinkled; stamens small; pistils small in the upper flower. Stalks stout, stiff; flowers upright. Flowering midseason.

Plant medium height, erect. Branches stout. Leaves long, medium-sized, stiff; leaflets long ovate, acuminate at the apex. Growth vigorous, but flowers relatively few. Bred by Zhaolou Peony Garden, Heze in 1970.

cv. Lu Yun Fu Ri

Hundred proliferate form, sometimes chrysanthemum form occurs. Flowers 16cm × 9cm, purplish red (61-C). The outer petals 5-6-whorled, gradually smaller from the outer to the inner petals, wrinkled; stamens normal; pistils developed, mixed purple and green coloured petals in the lower flower. Petals large, wrinkled; stamens normal; pistils normal, 3; floral discs white. Stalks straight, flowers upright. Flowering midseason, each individual flower is long-lasting.

Plant tall, erect. Branches stout. Leaves orbicular, medium-sized, stiff; leaflets long ovate, flat, acute at the apex. Growth vigorous, flowers many. Bred by Wangcheng Park, Luoyang in 1995.

Lü Yun Fu Ri

Anemone Form

2 – 3 outer whorls of wide and straight petals. Stamens completely petaloid and have become narrow and straight petals. Pistils normal or reduced.

Dan Ou Si

Zi Tuo Gui

Chu E Huang

cv. Dan Ou Si

Anemone form. Flowers 17cm × 6cm, pinkish white faintly bluish (65-D); the outer petals broad and large, the margins toothed, suffused with pale purple at the base, thin, stiff; the inner petals narrow, long, curved, sparse, with purple streaks on the centre; some petaloid filaments bearing anthers are present, a few stamens grow among petals; pistils small. Stalks slightly short; flowers upright. Flowering late midseason.

Plant dwarf, spreading. Branches stout. Leaves orbicular, medium-sized; leaflets nearly orbicular, obtuse at the apex. Growth fairly vigorous, flowers many. Bred in Heze in 1974.

cv. Zi Tuo Gui

Anemone or crown form. Flower buds orbicular; flowers 16cm × 7cm, purple (68-B); the outer petals 2-whorled, broadly ovate, suffused with deeper colour at the base; stamens petaloid, some petaloid filaments bearing anthers are present; pistils developed, petaloid, greenish coloured. Stalks fairly short, flowers growing at the same level with the leaves. Flowering midseason.

Plant fairly tall, spreading. Leaves long, small-sized; leaflets long ovate or long elliptic, acuminate at the apex, the margins slightly curved upwards. Growth vigorous, flowers many. Bred by Wangcheng Park, Luoyang in 1969.

cv. Chu E Huang

Anemone form, sometimes crown form occurs. Flowers 15cm × 6cm, light yellow suffused with pink (158-C); the outer petals 2-3-whorled, large, stiff, purple at the base; the inner petals narrowly long, soft, sparse, the some petaloid filaments bearing anthers are present, a few stamens grow among petals; pistils slightly petaloid or smaller, occasionally fertile. Stalks slender, long, flowers upright. Flowering midseason.

Plant tall, erect. Branches stout, stiff. Leaves orbicular, medium-sized, stiff, thick, sparse; leaflets ovate, acute at the apex. Growth vigorous, flowers many. Bred by Team 9, Zhaolou, Heze in 1966.

Qing Long Wo Mo Chi

cv. Qing Long Wo Mo Chi

Anemone form, sometimes crown form occurs. Flowers 19cm × 6cm, faintly dark purple (187-D); petals 2-whorled, broad and large, slightly curved upwards, suffused with dark purple; the inner petals twisted, normal stamens grow among inner petals; pistils developed, petaloid, greenish coloured. Stalks fairly short, slightly pliable, flowers lateral. Flowering midseason.

Plant medium height, spreading. Branches curved. Leaves orbicular, large-sized, thick; leaflets ovate, obtuse at the apex, pendulous. Growth fairly vigorous, flowers many. Classic variety.

Xian E

Zi Jing Hong

Shu Nü Zhuang

Qi Hua Xian Cai

San Bian Sai Yu

Fen Pan Tuo Gui

cv. Xian E

Anemone form. Flowers 14cm × 6cm, pink (65-A); the outer petals irregularly toothed, purplish red basal blotches; the inner petals narrow, sparse; a few stamens grow among the petals; pistils small; floral discs pale purple. Stalks slender, stiff, long; flowers upright. Flowering midseason.

Plant tall, erect. Branches slender, stiff. Leaves long, small-sized, sparse; leaflets ovate, apex acute at the apex, suffused with purplish red on the margins. Growth medium, flowers many. Bred by Zhaolou Peony Garden, Heze in 1989.

cv. Shu Nu Zhuang

Anemone form, sometimes single form occurs. Flowers 18cm × 6cm, pink (62-D); the outer petals 2-whorled, large, flat, nearly entire and regular, suffused with deep pink; the inner petals very small, twisted, sparse, some stamens normal; pistils small or slightly petaloid. Stalks fairly long, slightly pliable, flowers lateral. Flowering midseason.

Plant medium height, spreading. Branches fairly stout. Leaves long, medium-sized, thick, soft, fairly sparse; leaflets long ovate, acuminate at the apex, suffused with pale purple on the margins. Growth vigorous, flowers many. Bred by Zhaolou Peony Garden, Heze in 1967.

cv. San Bian Sai Yu

Anemone form. Flowers 15cm × 6cm, bud light green, at first pinkish white, later white (155-D); the outer petals fairly large, soft, toothed, purplish red basal blotches; the inner petals narrow, rugose, sparse, irregular; the some petaloid filaments bearing anthers are present, some stamens grows among petals; pistils small or developed, petaloid, greenish coloured. Stalks extremely short, flowers hidden among the leaves. Flowering midseason.

Plant medium height, erect. Branches slender, stiff. Leaves long, medium-sized, soft, extremely sparse; leaflets long ovate, acute at the apex, pendulous. Growth medium, flowers many. Bred by Zhaolou Peony Garden, Heze in 1965.

cv. Zi Jing Hong

Anemone form. Flowers 17cm × 6cm, purplish red (53-B); the outer petals 4-5-whorled, entire, large, with purplish red small streaks at the base; stamens partially developed, petaloid, small, toothed at the apex, curved inwards; the outer petals arranged closely, the majority of stamens normal in the flower centre, pistils normal; ovaries 12-18, floral discs smaller and absent. Stalks fairly short, flowers upright. Flowering midseason.

Plant dwarf, partially spreading. Branches slender, stiff. Leaves long, medium-sized; leaflets long ovate, acuminate at the apex. Growth medium, flowers many. Bred by Zhaolou Peony Garden, Heze in 1979.

cv. Qi Hua Xian Cai

Anemone form, sometimes crown form occurs. Flowers 18cm × 7cm, pink faintly bluish (65-D); the outer petals 2-whorled, with translucent streaks, the margins toothed, purple at the base; the inner petals narrowly long and curved, with purplish bluish streaks, anthers are usually present at the apex; pistils normal or small; Stalks slightly pliable, flowers slightly lateral. Flowering midseason.

Plant dwarf, partially spreading. Branches fairly slender, pliable. Leaves orbicular, large-sized, crowded, pendulous; leaflets broadly ovate, obtuse at the apex. Growth medium, flowers many; resistant to leaf spot. Bred by Zhaolou Peony Garden, Heze in 1963.

cv. Fen Pan Tuo Gui

Anemone form, sometimes crown form occurs. Flowers 14cm × 5cm, the outer petals pinkish purple (64-D), the inner petals light pinkish purple; the outer petals 2-3-whorled, broad and large, the inner petals small, wrinkled; stamens partially petaloid, some normal stamens grow among these petals; pistils usually developed, petaloid, greenish coloured. Stalks straight, flowers upright. Flowering late midseason.

Plant tall, erect. Branches fairly stout. Leaves orbicular, medium-sized, sparse; leaflets ovate. Growth medium, flowers many. Bred by Wangcheng Park, Luoyang in 1969.

cv. Man Mian Chun

Anemone form, sometimes crown form occurs. Flowers 20cm × 10cm, pink (62-C); the outer petals 3-4-whorled, broad and large, gradually smaller from the outer to the inner whorls, some petals toothed on the margins, the inner petals very small, wrinkled, with deep red streaks at the base; stamens partially normal or only filaments developed into twisted petals; some petaloid filaments bearing anthers are present, pistils deformed; floral discs greenish white. Stalks straight, purplish red, flowers upright. Flowering late.

Plant tall, erect. Branches fairly stout. Leaves orbicular, large-sized; leaflets broadly ovate, obtuse at the apex. Growth vigorous, flowers normal. Bred by Wangcheng Park, Luoyang in 1995.

Man Mian Chun

GOLDEN-CIRCLE FORM

2 – 3 outer whorls of wide and large petals. Most stamens petaloid, but a whorl of normal stamens remains as a golden circle between the narrow interior petals and the wide outer petals. Pistils normal, or petaloid, or reduced.

Fen Mian Tao Hua

cv. Fen Mian Tao Hua

Golden circle form. Flowers 20cm × 7 cm, deep pinkish red (73-B); the outer petals 6-whorled, broad and large, slightly toothed, thick, suffused with light red at the base, apex pinkish white; the inner petals fairly large, wrinkled, a distinct whorl of normal stamens grow between outer and inner petals; pistils developed, petaloid, greenish coloured. Stalks stout, stiff, flowers lateral. Flowering late midseason.

Plant tall, partially spreading. Branches stout. Leaves orbicular, medium-sized, thick, stiff, crowded; leaflets ovate with numerous lobes, apex acute, margins slightly curved upwards, upper surface green. Growth vigorous, flowers many. Bred by Zhaolou Peony Garden, Heze in 1983.

cv. Yu Mei Ren

Golden circle form. Flowers 18cm × 7cm, pinkish white (49-D), soft; the outer petals 3-whorled, large, thin, fairly stiff, suffused with light pink at the base; the inner petals fairly large, a distinct whorl of normal stamens grow between outer and inner petals; stigmas smaller, floral discs pinkish red. Stalks fairly pliable, flowers lateral. Flowering late.

Plant medium to dwarf, partially spreading. Branches fairly slender. Leaves long, medium-sized; petiolules long; leaflets long elliptic or long ovate, acuminate at the apex, the margins curved upwards. Winter buds, new branches, leaves, petioles light green, high ornamental value in early spring. Growth medium, flowers many. Bred by Zhaolou Peony Garden, Heze in 1986.

Yu Mei Ren

CROWN FORM

Wide and expanded outer petals. Completely petaloid stamens usually with the appearance of becoming larger from outside to inside, sometimes mixed with a few narrow silk-like incompletely petaloid stamens. Pistils are petaloid, reduced or completely disappeared. The centre of the flower is raised, forming a crown shape.

cv. Kun Shan Ye Guang

Crown form. Flowers 16cm × 6cm, white (155-D), sparkling and tidy; the outer petals 3-4-whorled, stiff, flat, slightly suffused with pale purple at the base; the inner petals large, wavy, stamens completely petaloid; pistils developed, petaloid, greenish coloured. Stalks fairly short, flowers usually hidden among leaves. Flowering late.

Plant medium height, spreading. Branches stout. Leaves orbicular, medium-sized, stiff; leaflets ovate, acute at the apex. Growth vigorous; buds damaged by spring frost, flowers normal. Classic variety.

Kun Shan Ye Guang

Kun Shan Ye Guang

cv. Xue Li Zi Yu

Crown form. Flowers 16cm × 8cm, white faintly bluish (155-D); the outer petals 2-whorled, large, flat, deep purple basal blotches, the inner petals large, some petaloid filaments bearing anthers are present; pistils developed, petaloid, greenish coloured. Stalks short, flowers upright. Flowering late midseason.

Plant dwarf, erect. Branches stiff. Leaves long, medium-sized, soft, crowded; leaflets elliptic or long ovate, acute at the apex. Growing medium, flowers many; resistant to leaf spot. Bred by Zhaolou Peony Garden, Heze in 1972.

cv. Li Hua Xue

Crown form. Flowers 16cm × 7cm, white (155-D); the outer petals 2-whorled, large, soft, with translucent streaks; the inner petals crowded, rugose, a few stamens grow among the petals; pistils developed, petaloid, greenish coloured. Stalks slender, long, slightly pliable, light green, flowers lateral. Flowering midseason.

Plant dwarf, spreading. Branches slender. Leaves long, small-sized, sparse, ternate pinnates; leaflets ovate, acute at the apex, the margins curved downwards. Growth weak, flowers many, difficult to propagate. Classic variety.

cv. Jin Si Guan Ding

Crown form. Flowers 18cm × 10cm, white (158-D); the outer petals 3-whorled, broad and large, entire and regular, stiff, suffused with purple at the base; the inner petals arranged closely, some petaloid filaments bearing anthers are present; pistils small. Stalks stout, straight, flowers upright. Flowering midseason.

Plant tall, partially spreading. Branches stout. Leaves orbicular, large-sized; leaflets long elliptic, apex obtuse, the margins slightly twisted; terminal leaflets deeply lobed. Growth vigorous, flowers many. Bred by Wangcheng Park, Luoyang in 1969.

Xue Li Zi Yu

Li Hua Xue

Jin Si Guan Ding

cv. Yu Gu Bing Ji

Crown form. Flowers 14cm × 8cm, white (158-C); the outer petals 2-whorled, broad and large, entire and regular, the inner petals very small, erect, nearly all stamens petaloid; pistils normal; floral discs white. Stalks straight, flowers upright. Flowering early.

Plant dwarf, erect. Branches slender, stiff. Leaves orbicular, medium-sized; leaflets ovate, obtuse at the apex, the margins curved upwards. Growth medium, flowers many. Bred by Wangcheng Park, Luoyang in 1969.

cv. Jin Xing Xue Lang

Crown form, sometimes anemone form occurs. Flowers 20cm × 9cm, white (158-D); the outer petals 2-whorled, large, stiff, flat, purplish red basal blotches; the inner petals narrowly long, wrinkled, slightly crowded, some petaloid filaments bearing anthers are present, a few stamens grow among petals, pistils slightly petaloid, floral discs pale purple. Stalks fairly short, pliable, flowers lateral. Flowering late.

Plant medium height, spreading. Branches slender, pliable, curved. Leaves orbicular, medium-sized, slightly sparse; leaflets ovate, obtuse at the apex, the margins slightly wavy. Growth medium, flowers normal. Bred by Heze Flower Garden in 1990.

cv. Yin Gui Piao Xiang

Crown form, sometimes anemone or single form occurs. Flowers 22cm × 8cm, white (155-D); the outer petals 2-whorled, large, thin, translucent, suffused with pinkish white at the base; the inner petals sparse, slightly curved over, some petaloid filaments bearing anthers are present, a few stamens grow among the petals; pistils slightly petaloid, floral discs pale purple. Stalks stout, short, flowers upright. Flowering late midseason.

Plant medium height, erect. Branches stout. Leaves orbicular, medium-sized, slightly thick, fairly crowded; leaflets ovate, apex acute. Growth vigorous, flowers many; tolerant of adverse conditions. Bred by Zhaolou Peony Garden, Heze in 1995.

Yu Gu Bing Ji

Jin Xing Xue Lang

cv. Song Bai

Crown form. Flowers 16cm × 8cm, white (155-D); the outer petals 2-whorled, large, flat, apex toothed, suffused with pinkish red at the base; the inner petals very small, wrinkled, crowded, some petaloid filaments bearing anthers are present at the petal apex, some stamens grow among the petals; pistils slightly petaloid. Stalks slightly short, flowers upright or lateral. Flowering late midseason.

Plant dwarf, spreading. Branches slightly curved. Leaves long, medium-sized, sparse; leaflets long ovate, obtuse at the apex, the margins slightly curved upwards, terminal leaflets pendulous. Growth medium, flowers many. Classic variety.

Crown Form 141

Yin Gui Piao Xiang

Song Bai

cv. Jing Gu Chun Qing

Crown form. Flowers 15cm × 7cm, pinkish white faintly bluish (73-D), lustrous; the outer petals 2-3 whorled, fairly large, shallowly lobed at the apex, the inner petals very small, erect, suffused with purple at the base; pistils normal, stigmas light red; floral discs purplish red; stalks erect, flowers upright. Flowering midseason.

Plant dwarf, erect. Branches fairly slender, stiff. Leaves orbicular, medium-sized, broad and large, thick; leaflets crowded, ovate, obtuse at the apex. Growth vigorous, flowers many. Bred by Guose Peony Garden, Luoyang in 1992.

cv. Qing Shan Guan Xue

Crown form. Flowers 12cm × 5cm, at first pinkish white, later white (155-D), the outer petals 2-whorled, stiff, apex irregularly toothed, suffused with pale purple at the base; the inner petals sparse, apex toothed; pistils developed, petaloid, greenish coloured. Stalks short, flowers upright or lateral. Flowering early.

Plant dwarf, partially spreading. Branches stout. Leaves long, small-sized; leaflets long ovate, shallowly lobed, obtuse at the apex. Growth vigorous, flowers relatively few, irregular in form. Classic variety.

cv. Bai Lian Xiang

Crown form, sometimes lotus form or anemone form occurs. Flowers 15cm × 5cm, at first pinkish white, later white (155-D); the outer petals large, stiff, purplish red basal blotches, centre parts of the inner petals wrinkled, small, enlarged near apex, some petaloid filaments bearing anthers are present, a few stamens grow among petals, pistils small or slightly petaloid. Stalks long, stiff, flowers upright. Flowering early.

Plant medium height, erect. Branches fairly stout. Leaves long, medium-sized; leaflets ovate, deeply lobed, acuminate at the apex, the margins slightly curved upwards. Growth vigorous; flowers many, varied in form, several flower forms usually found on a plant; tolerant of diseases and salinity-alkalinity. Bred by Zhaolou Peony Garden, Heze in 1976.

cv. Xue Gui

Crown form, sometimes anemone form occurs. Flowers 15cm × 7cm, white (155-D); the outer petals 2-whorled, large, flat, irregularly toothed, suffused with light red at the base; the inner petals fairly soft, spare, rugose, some petaloid filaments bearing anthers are present, a few stamens grow among the petals; pistils small or developed, petaloid, greenish coloured. Stalks short, flowers lateral. Flowering midseason.

Plant medium height, partially spreading. Branches stout, stiff. Leaves orbicular, small-sized, sparse; leaflets ovate, apex acute. Growth vigorous, flowers many; resistant to leaf spot. Bred by Team 9, Zhaolou, Heze in 1974.

142 CHINESE TREE PEONY

Jin Gu Chun Qing

Bai Lian Xiang

Qing Shan Guan Xue

Xue Gui

pendulous, the margins curved upwards. Growth fairly vigorous, flowers relatively few. Bred by Zhaolou Peony Garden, Heze in 1963.

cv. Bing Hu Xian Yu

Crown form. Flowers 15cm × 7cm, at first light pinkish white, later white faintly bluish (155-D); the outer petals 2-whorled, large, flat, suffused with purple at the base; the inner petals wrinkled, of a similar size, regular, with light purple streaks in petal centre; pistils small or slightly petaloid. Stalks fairly long, pliable, flowers lateral. Flowering midseason.

Plant medium height, spreading. Branches relatively slender, pliable, curved. Leaves orbicular, medium, thick; leaflets ovate, the margins curved upwards, apex obtuse or acute. Growth medium, flowers many. Bred by Team 9, Zhaolou, Heze in 1968.

cv. Huang Cui Yu

Crown form. Flowers 16cm × 9cm, yellowish (158-C); the outer petals 2-3-whorled, broad and large, reflexed, suffused with light red at the base; a whorl of incompletely petaloid stamens are present between the outer and the inner petals, very small, some petaloid filaments bearing anthers are present, light green; pistils developed, petaloid, greenish coloured. Stalks fairly long, stiff, yellowish green, flowers upright or lateral. Flowering midseason.

Plant medium height, partially spreading. Branches stout. Leaves orbicular, large-sized, slightly sparse; leaflets ovate or broadly acute, the margins slightly curved upwards, apex obtuse. Growth vigorous, flowers many; resistant to diseases. Bred by Zhaolou Peony Garden, Heze in 1983.

cv. Fen Qing Shan

Crown form. Flowers 14cm × 8cm, pinkish white faintly bluish (69-D); the outer petals thin, suffused with pale purple at the base; the inner petals crowded, regular, some petaloid filaments bearing anthers are present; pistils developed, petaloid, greenish coloured. Stalks short, flowers upright. Flowering midseason.

Plant medium height, partially spreading. Branches fairly stout. Leaves long, medium-sized, soft, crowded; leaflets long ovate, acuminate at the apex,

Huang Cui Yu

Crown Form 143

Fen Qing Shan

Wu Xia Mei Yu

cv. Bai Yu

Crown form. Flowers 16cm × 8cm, at first pinkish white, later white (155-D), the outer petals 2-whorled, flat, suffused with purple at the base; the inner petals smaller, soft, wrinkled, regular, crowded, some petaloid filaments bearing anthers are present; pistils smaller, or developed, petaloid, yellowish green coloured. Stalks slightly long, flowers lateral. Flowering midseason.

Plant dwarf, partially spreading. Branches slightly slender. Leaves orbicular, medium-sized, thick, sparse; leaflets ovate with some and shallow lobes, obtuse at the apex, upper surface rough, green, suffused with purple. Growth fairly vigorous; flowers many. Classic variety.

Bing Hu Xian Yu

Xue Ta

cv. Wu Xia Mei Yu

Crown form. Flowers 16cm × 6cm, white (155-D); the outer petals 2-whorled, thin, soft, suffused with pale purple at the base; the inner petals soft, sparse; some petaloid filaments bearing anthers are present, a few stamens grow among petals; pistils developed, petaloid, greenish coloured. Stalks fairly pliable, flowers laterally pendulous. Flowering midseason.

Plant dwarf, spreading. Branches fairly slender, curved. Leaves orbicular, medium-sized; leaflets ovate, thick, obtuse at the apex, the margins curved upwards. Growth slow. Bred by Zhaolou Peony Garden, Heze in 1983.

cv. Xue Ta

Crown form, sometimes lotus form or anemone form occurs. Flowers 16cm × 9cm, white faintly orange (155-D); the outer petals 2-whorled, large, flat, suffused with pinkish red at the base; the inner petals crowded, regular, some petals curved over; a few stamens grow among petals; pistils small or petaloid. Stalks stout, flowers upright. Flowering midseason.

Plant medium height, partially spreading. Branches stout. Leaves long, large-sized, crowded, thick; leaflets long ovate, the margins curved upwards, acuminate at the apex. Growth vigorous, flowers many. Bred by Team 9. Zhaolou, Heze in 1973.

Bai Yu

Jing Yu

Jing Yu

cv. Jing Yu

Crown form. Flowers 17cm × 9cm, at first pinkish white, later white (155-D); the outer petals 2-whorled, large, thin, flat, suffused with pinkish red at the base; the inner petals narrowly long, wrinkled, regular and tidy, apex usually shallowly toothed, a few stamens grow among petals; stigmas smaller; floral discs purple. Stalks slender, stiff, long, flowers upright. Flowering early.

Plant tall, erect. Branches slender, stiff. Leaves long, medium-sized, soft, sparse; leaflets elliptic, apex acute. Growth vigorous; flowers profuse; tolerant of adverse conditions. Bred by Sun Jingyu from Heze Flower Garden, Heze in 1978.

Wa Wa Mian

Crown Form

Xing Hua Bai

Gu Tong Yan

Jin Lun Huang

Bing Zhao Lan Yu

cv. Wa Wa Mian
Crown form. Flowers 15cm × 6cm, pinkish white (56-C); the outer petals 2-whorled, large; the inner petals rugose, crowded; some stamens grow among petals, pistils petaloid or smaller. Stalks stout, flowers upright or lateral. Flowering midseason.
Plant dwarf, spreading. Branches stout, stiff. Leaves orbicular, medium-sized, thick; leaflets ovate with some lobes, apex acute. Growth medium, flowers many. Bred by Zhaolou Peony Garden, Heze in 1978.

cv. Xing Hua Bai
Crown form. Flowers 16cm × 7cm, at first pinkish white, later pure white (155-D); the outer petals 1-2-whorled, usually toothed, thin, soft, with pink at the base, all stamens petaloid; pistils small. Stalks fairly stiff, flowers upright or lateral. Flowering midseason.
Plant tall, partially spreading. Branches fairly slender, stiff. Leaves long, medium-sized, sparse, thin, stiff; petiolules long; leaflets long ovate, acuminate at the apex. Growth fairly vigorous, flowers many. Bred by Zhaolou Peony Garden, Heze in 1989.

cv. Jin Lun Huang
Crown form, sometimes anemone form occurs. Flowers 16cm × 5cm, yellowish (8-D); the outer petals 2-whorled, large, with translucent streaks, toothed on the margins, purple at the base; the inner petals narrowly long, fairly stiff, some stamens grow among the petals; pistils small. Stalks slender, straight, flowers upright. Flowering midseason.
Plant medium height, erect. Branches slender, stiff. Leaves orbicular, small-sized, sparse; leaflets ovate, apex acute, the margins slightly curved upwards; tomentose beneath. Growth weak, flowers many, difficult to propagate. Classic variety.

cv. Bing Zhao Lan Yu
Crown form. Flowers 13cm × 7cm, pinkish white faintly bluish (76-D); the outer petals 2-whorled, large, with tiny round teeth, suffused with purple at the base, the inner petals straight; pistils petaloid. Stalks fairly long, stiff, flowers lateral. Flowering early.
Plant medium height, partially spreading. Branches fairly stout. Leaves orbicular, medium-sized; leaflets ovate, apex acute, pendulous, the margins curved upwards. Growth slow, flowers many. Bred by Zhaolou Peony Garden, Heze in 1969.

cv. Gu Tong Yan
Crown form. Flowers 15cm × 6cm, light yellow faintly brown (29-D); the outer petals 2-whorled, large, stiff, dark purple basal blotches, the inner petals wrinkled, regular, some stamens grow among the petals; pistils small or slightly petaloid. Stalks long, stiff, flowers upright. Flowering midseason.
Plant medium height, erect. Branches slender, stiff. Leaves long, small-sized, sparse; leaflets long ovate, acuminate at the apex, the margins curved upwards. Growth weak; flowers many, leaves falling early. Bred by Team 9, Zhaolou, Heze in 1980.

Jin Yu Xi

Dou Lü

Can Xue

Qing Xiang Bai

cv. Dou Lu
Crown or globular form. Flowers 12cm × 6cm, yellowish green (144-D); petals 2-3-whorled, thick, stiff, purple basal blotches; the inner petals crowded, rugose; pistils petaloid or small. Stalks slender, pliable, flowers pendulous. Flowering late.

Plant dwarf, spreading. Branches relatively slender. Winter buds narrowly conical, aquiline, pale brownish green, with red at the apex of bud scales. Leaves long, medium-sized; leaflets broadly ovate, apex acute, pendulous, green slightly suffused with purple above, dense pilose beneath. Growth medium, flowers many. Classic variety.

cv. Can Xue
Crown form, sometimes anemone form occurs. Flowers 16cm × 7cm, at first creamy yellow, later white(158-C); the outer petals 2-whorled, large, stiff, nearly entire and regular, suffused with purple at the base; the inner petals rugose, crowded, some petaloid filaments bearing anthers are present; pistils small or developed, petaloid, yellowish green coloured. Stalks long, flowers upright. Flowering midseason.

Plant medium height, erect. Branches slender, stiff. Leaves orbicular, medium-sized, thick, sparse; leaflets ovate, with some and shallow lobes, obtuse at the apex. Growth vigorous, flowers relatively few. Bred by Zhaolou Peony Garden, Heze in 1989.

cv. Qing Xiang Bai
Crown form, sometimes lotus or anemone form occurs. Flowers 15cm × 7cm, at first creamy yellow, later white (155-D); the outer petals 2-3-whorled, large, stiff, rarely suffused with pink at the base; the inner petals wrinkled; some petaloid filaments bearing anthers are present, some stamens grow among petals; pistils small or partially petaloid. Stalks slender and long, stiff, flowers upright. Flowering early.

Plant tall, erect. Branches slender, stiff. Leaves orbicular, medium-sized, thick, relatively sparse; leaflets ovate, apex acute, deep green, the margins ruffled. Growth vigorous, flowers many, several flower forms found on a plant. Bred by Zhaolou Peony Garden, Heze in 1975.

cv. Jin Yu Xi
Crown form. Flowers 14cm × 7cm, white faintly yellow (158-D); the outer petals 2-whorled, large, thin, suffused with pale purple at the base; the inner petals crowded, rugose, some petaloid filaments bearing anthers are present; pistils small. Stalks slender, stiff, straight, flowers lateral. Flowering late midseason.

Plant dwarf, partially spreading. Branches slender. Leaves long, medium-sized, sparse; leaflets ovate, acuminate at the apex. Growth weak, flowers many, neatform. Classic variety.

cv. Bi Hai Fo Ge

Crown form. Flowers 18cm × 8cm, pinkish white faintly bluish (65-B); the outer petals 3-4-whorled, the inner petals fairly large, crowded, full; pistils developed, petaloid, greenish coloured. Stalks stout, relative pliable, flowers lateral. Flowering late.

Plant medium height, partially spreading. Branches stout. Leaves orbicular, large-sized, large, thick; leaflets orbicular with numerous lobes, obtuse at the apex, the margins curved upwards. Growth extremely vigorous, flowers normal. Bred by Jingshan Park, Beijing in 1995.

cv. Fen Pan Jin Qiu

Crown form. Flowers 16cm × 6cm, pink (38-C); the outer petals 2-whorled, broad and large, curved downwards; the inner petals crowded, arranged regularly; pistils small or absent. Stalks stout, stiff; flowers lateral. Flowering late.

Plant medium height, spreading. Branches relatively stout. Leaves long, medium-sized; petiolules very long, curved downwards, leaflets pendulous, long ovate, with numerous lobes and slightly curved upwards on the margins. Growth vigorous, flowers relatively few. Bred by Zhaolou Peony Garden, Heze in 1968.

Fen Pan Jin Qiu

Bi Hai Fo Ge

Xi Gua Rang

cv. Xi Gua Rang

Crown form. Flowers 16cm × 8cm, red (52-C), apex pink; the outer petals 2-3-whorled, lighter colour, red at the base; the inner petals regular and neat, arranged closely; pistils petaloid or short. Stalks fairly short, fairly pliable, flowers lateral. Flowering midseason.

Plant dwarf, partially spreading. Branches slender, fairly pliable. Leaves long, medium-sized, fairly sparse; leaflets long ovate-lanceolate, acuminate at the apex. Growth medium; flowers many, neatform. Bred by Zhaolou Peony Garden, Heze in 1978.

cv. Yao Huang

Crown form, sometimes golden circle form occurs. Flowers 16cm × 10cm, yellowish (8-D); the outer petals 3-4-whorled, stiff, purple basal blotches, the inner petals wrinkled, crowded, some petaloid filaments bearing anthers are present, pistils small or petaloid. Stalks long, straight, flowers upright. Flowering midseason.

Plant tall, erect. Branches relatively slender, stiff. Leaves orbicular, medium-sized; leaflets ovate with some lobes, apex obtuse, terminal leaflets pendulous. Growth fairly vigorous; flowers many, normal, neatform. Classic variety.

Yao Huang

Lan Tian Yu

Lu Fen

cv. Jin Yu Jiao Zhang

Crown form, sometimes anemone form occurs. Flowers 15cm × 6cm, at first yellow, later creamy white (155-D); the outer petals 2-whorled, large, stiff, nearly entire and regular, with only tiny teeth in the centre of apex, suffused with pinkish red at the base, the inner petals wrinkled, slightly curved over, some petaloid filaments bearing anthers are present; pistils small or petaloid. Stalks short, flowers upright. Flowering early.

Plant dwarf, partially spreading. Branches fairly stout. Leaves orbicular, medium-sized, stiff, crowded; leaflets ovate, apex obtuse, the margins curved upwards; upper surface rough, yellowish green. Growth vigorous, flowers relatively few. Classic variety.

cv. Lan Tian Yu

Crown form. Flowers 15cm × 5cm, pink faintly bluish (65-D); outer petals 2-whorled, large, flat, with faintly purple streaks, suffused with purple at the base; the inner petals twisted, crowded, some petaloid filaments bearing anthers are present; pistils small or developed, petaloid, greenish coloured. Stalks fairly short, flowers upright. Flowering late midseason.

Plant dwarf, partially spreading. Branches relatively stout. Leaves orbicular, medium-sized, thick; leaflets ovate, apex obtuse. Flowers many, neatform. Classic variety.

cv. Lu Fen

Crown form. Flowers 17cm × 7cm, pink (38-D); the outer petals 2-whorled, thick, apex toothed, suffused with pinkish red at the base; the inner petals wrinkled, some petaloid filaments bearing anthers are present, some stamens grow among the petals; pistils developed, petaloid, greenish coloured. Stalks stiff, flowers upright. Flowering early.

Plant medium height, partially spreading. Leaves long, medium-sized, thick, stiff, flat; leaflets ovate or long elliptic, apex acute. Growth vigorous; flowers many, neatform, plants elegant. Bred by Zhaolou Peony Garden, Heze in 1969.

cv. Fen Zhong Guan

Crown form. Flowers 16cm × 9cm, pink (38-D); the outer petals 2-3-whorled, large, suffused with pinkish red at the base; the inner petals rugose, crowded, regular, some curved over; pistils developed, petaloid, yellowish green. Stalks short, stiff, flowers upright. Flowering midseason.

Plant medium height, spreading. Branches fairly stiff. Leaves long, medium-sized, slightly soft, crowded; leaflets long ovate, acuminate at the apex. Growth vigorous; flowers many, neatform; tolerant of diseases. Well-known pink cultivar. Bred by Zhaolou Peony Garden, Heze in 1973.

Jin Yu Jiao Zhang

Fen Zhong Guan

Crown Form

cv. Bai He Wo Xue

Crown form, sometimes anemone form occurs. Flowers 17cm × 6cm, at first pink, later white faintly bluish (155-D); the outer petals 2-whorled, broad and large, irregularly toothed at the apex; the inner petals very small, crowded, wrinkled, some petaloid filaments bearing anthers are present, a few stamens grow among the petals; pistils small. Stalks fairly long, flowers upright. Flowering midseason.

Plant medium height, spreading. Branches relatively stout. Leaves orbicular, medium-sized, sparse; leaflets ovate, apex obtuse or acute, pendulous, the margins slightly curved upwards, suffused with purple. Growth vigorous, flowers many. Bred by Zhaolou Peony Garden, Heze in 1969.

cv. Bing Ling Zi

Crown form. Flowers 16cm × 6cm, pink (38-C); the outer petals 2-3-whorled, large, thin, soft, apex usually toothed, purplish red basal blotches, the inner petals very small, wrinkled, some petaloid filaments bearing anthers are present, some stamens grow among the petals; pistils developed, petaloid, yellowish green. Stalks short, flowers hidden among leaves. Flowering midseason.

Plant medium height, spreading. Branches curved. Leaves long, medium-sized, sparse; leaflets elliptic, the margins curved upwards, acute at the apex. Growth vigorous, flowers relatively few. Bred by Zhaolou Peony Garden, Heze in 1964.

Fen Lan Lou

Yu Xi Ying Yue

cv. Fen Lan Lou

Crown form. Flowers 16cm × 9cm, pink faintly bluish (73-C); the outer petals 2-3-whorled, large, thin, suffused with purple at the base, the inner petals wrinkled, relatively sparse; pistils developed, petaloid, greenish coloured. Stalks slightly short, flowers lateral. Flowering midseason.

Plant medium height, partially spreading. Branches relatively stout. Leaves long, large-sized, thick, soft; leaflets ovate lanceolate, acuminate at the apex, pendulous, the margins curved upwards. Growth vigorous, flowers relatively few. Bred by Zhaolou Peony Garden, Heze in 1968.

cv. Yu Xi Ying Yue

Crown form, sometimes anemone form occurs. Flowers 15cm × 6cm, yellowish (158-C); the outer petals 2-whorled, large, stiff, purple basal blotches, the inner petals wrinkled, sparse; some petaloid filaments bearing anthers are present, a few stamens grow among the petals, pistils slightly petaloid. Stalks short, stiff, flowers upright. Flowering midseason.

Plant medium height, partially spreading. Branches stiff. Leaves orbicular, medium-sized, crowded; leaflets ovate with some lobes, obtuse at the apex, the margins curved upwards. Flowers many, very adaptable. Bred by Zhaolou Peony Garden, Heze in 1975.

Bai He Wo Xue

Bing Ling Zi

Dou Zhu

Luo Yang Chun

Jiu Tian Lan Yue

cv. Dou Zhu

Crown form. Flowers 15cm × 9cm, pink faintly bluish (65-C); the outer petals 2-3-whorled, broad and large, with a thick purple streak along petal centre, dark purple basal blotches; the inner petals crowded, usually toothed at the apex; nearly all stamens petaloid, pistils small. Stalks slightly pliable. Flowering midseason.

Plant dwarf, spreading. Branches slender, curved. Leaves long, medium-sized; leaflets ovate or long ovate, acute at the apex, pendulous, the margins curved upwards. Growth weak, flowers relatively few, sprouts twice each year, in summer and autumn. Classic variety.

cv. Luo Yang Chun

Crown form. Flowers 17cm × 9cm, bluish pink (62-C); the outer petals 2-whorled, large, with a deep tooth at the apex, the inner petals fairly small, layered, all stamens petaloid, some petaloid filaments bearing anthers are present; pistils 3, normal, stigmas red; floral discs purplish red. Stalks fairly straight, flowers upright. Flowering midseason.

Plant medium height, erect. Branches fairly stout. Leaves orbicular, medium-sized, sparse, thick; petiolules long; leaflets ovate, flat, obtuse at the apex. Growth vigorous, flowers many. Classic variety.

cv. Jiu Tian Lan Yue

Crown form, sometimes anemone form occurs. Flowers 23cm × 8cm, pink faintly bluish (62-D); the outer petals 2-3-whorled, broad and large, with numerous teeth, purplish red basal blotches; the inner petals narrowly long, wrinkled, some petaloid filaments bearing anthers are present, some stamens grow among the petals; pistils developed, petaloid, greenish coloured. Stalks long, stiff, pale purple, flowers upright. Flowering early midseason.

Plant tall, erect. Branches stiff. Leaves orbicular, medium-sized, stiff, sparse; leaflets ovate or long ovate, apex obtuse, pendulous. Growth vigorous, flowers many; tolerant of low temperature during bud stage. Bred by Heze Flower Garden in 1972.

Crown Form

Cui Mu

Gu Cheng Chun Se

E Mei Xian Zi

cv. Cui Mu

Crown form. Flowers 15cm × 7cm, pink faintly purple (62-D); the outer petals 2-whorled, large, stiff, with green at the back of the first whorl, deep purplish red basal blotches; the inner petals slightly sparse, twisted and wrinkled; stamens partially infertile; pistils developed, petaloid, tiny, greenish coloured. Stalks fairly pliable, flowers lateral. Flowering midseason.

Plant medium height, spreading. Branches somewhat curved. Leaves orbicular, medium-sized, soft, slightly sparse; leaflets ovate, obtuse at the apex, pendulous. Growth medium, flowers many. Bred by Team 9, Zhaolou, Heze, 1970.

cv. Gu Cheng Chun Se

Crown form. Flowers 17cm × 13cm, yellowish (27-D); the outer petals 2-whorled, stiff, large, deep purple basal blotches, radiating to the apex; the inner petals very small, rugose, crowded, some petaloid filaments bearing anthers are present, a few stamens grow among the petals; pistils small or petaloid. Stalks stiff, long, flowers upright. Flowering midseason.

Plant medium height, erect. Branches stiff. Leaves orbicular, medium-sized, sparse; leaflets ovate, obtuse at the apex. Growth medium; flowers relatively few, neatform. Bred by Zhaolou Peony Garden, Heze in 1985.

cv. E Mei Xian Zi

Crown form. Flowers 14cm × 11cm, pinkish white faintly bluish (62-C); petals 2-whorled, large, entire and regular, the inner petals erect, slightly wrinkled, deep pink at the base; stamens completely petaloid; pistils normal or occasionally petaloid, stigmas red; floral discs greenish white. Stalks pliable; flowers pendulous. Flowering late.

Plant medium height, partially spreading. Branches stout. Leaves long, medium-sized; leaflets long elliptic with numerous lobes, acuminate at the apex. Growth vigorous, flowers many. Bred by Wangcheng Park, Luoyang in 1969.

cv. Tian Xiang Zhan Lu

Crown form, sometimes anemone form occurs. Flowers 14cm × 6cm, pink faintly purple (73-C); the outer petals 2-whorled, large, soft, usually toothed, suffused with purple at the base; the inner petals narrowly long, soft, rugose, sparse; some petaloid filaments bearing anthers are present, some stamens grow among petals; pistils small or petaloid. Stalks long, stiff, flowers upright. Flowering midseason.

Plant high, erect. Branches slender, stiff. Leaves orbicular, medium-sized, thick, sparse; leaflets ovate, shallowly lobed, apex acute. Growth medium, flowers many. Bred by Zhaolou Peony Garden, Heze in 1963.

cv. Sheng Lan Lou

Crown form. Flowers 17cm × 6cm, pink faintly bluish (65-B); the outer petals 2-whorled, large, soft, suffused with purple at the base; the inner petals wrinkled; pistils petaloid. Stalks fairly short, stout, flowers upright. Flowering late.

Plant medium height, partially spreading. Branches stout. Leaves long, medium-sized, thick; leaflets ovate, apex acute, pendulous, the margins reflexed and twisted, upper surface green. Growth medium; flowers relatively few. Bred by Zhaolou Peony Garden, Heze in 1963.

cv. Yue Guang

Crown form. Flowers 18cm × 10cm, pink faintly bluish (56-D); the outer petals 3-whorled, large, flat, apex toothed, suffused with purple at the base; inner petals slightly large, fairly sparse, apex toothed, slightly wrinkled; stamens completely petaloid; pistils developed, petaloid, greenish coloured. Stalks stout, stiff, flowers upright. Flowering late.

Plant dwarf, partially spreading. Branches stout. Leaves long, large-sized, soft, thick; leaflets elliptic, acuminate at the apex, the margins slightly curved, upper surface deep green. Growth medium, flowers many; growth compact, tolerant of low temperature during bud stage. Bred by Heze Flower Garden in 1990.

Tian Xiang Zhan Lu

Sheng Lan Lou

Yue Guang

Hai Bo

cv. Hai Bo

Crown form, sometimes lotus or anemone form occurs. Flowers 16cm × 7cm, bluish pink (55-C); the outer petals 3-4-whorled, large, stiff, flat, suffused with red at the base; the inner petals wrinkled, large, sparse, a few stamens grow among the petals, pistils small or developed, petaloid, greenish coloured. Stalks slender, stiff, fairly long, flowers upright. Flowering early.

Plant medium height, partially spreading. Branches slender, stiff. Leaves orbicular, medium-sized, stiff, sparse; leaflets long ovate or long elliptic, obtuse at the apex. Growth medium; flowers profuse, several flower forms, each individual flower long-lasting, tolerant of strong sunshine. Bred by Heze Flower Garden in 1973.

cv. Zi Lan Kui

Zi Lan Kui

Crown form. Flowers 16cm × 8cm, bluish pink (73-C); the outer petals 2-whorled, large, the margins toothed shallowly, suffused with purple at the base; the inner petals crowded and regular, rugose, the terminal ones fairly large, some petaloid filaments bearing anthers are present; pistils small or petaloid. Stalks short, stiff, flowers upright. Flowering midseason.

Plant medium height, partially spreading. Branches stout. Leaves orbicular, medium-sized, thick, crowded; leaflets ovate, obtuse at the apex, the margins slightly curved upwards, upper surface rough, green. Growth vigorous; flowers many, flowers neatform; growth compact. Bred by Zhaolou Peony Garden, Heze in 1969.

cv. Ruan Zhi Lan

Crown form. Flowers 17cm × 7cm, pink faintly bluish (73-D); the outer petals 2-3-whorled, flat; the inner petals layered, usually toothed at the apex, suffused with purple at the base; pistils developed, petaloid, greenish coloured. Stalks short, pliable, flowers laterally pendulous. Flowering late.

Plant tall, spreading. Branches stout. Leaves long, large-sized, thick, soft; leaflets broadly ovate lanceolate, acuminate at the apex, pendulous. Growth vigorous, flowers relatively few, leaves falling late; resistant to leaf spot; high yield of danpi from roots for medicine. Bred by Zhaolou Peony Garden, Heze in 1965.

Ruan Zhi Lan

cv. Bai Hua Kui

Crown form, anemone or lotus form occurs. Flowers 18cm × 8cm, red (64-C); the outer petals 2-3-whorled, large, thin, suffused with red at the base, the inner petals very small, rugose, crowded, some petaloid filaments bearing anthers are present, a few stamens grow among the petals, pistils developed, petaloid, greenish coloured. Stalks stiff, pale purple, flowers upright. Flowering midseason.

Plant tall, partially spreading. Branches stiff. Leaves orbicular, large-sized, soft, thick; leaflets long ovate, obtuse at the apex, pendulous, upper surface rough, deep green, suffused with purple. Growth vigorous; flowers many, not tolerant of strong sunshine; tolerant of adverse conditions. Bred by Heze Flower Garden in 1981.

cv. Bai Hua Du

Crown form. Flowers 13cm × 5cm, light red (50-C), bright; the outer petals 2-3-whorled, suffused with deep purplish red at the base, the inner petals wrinkled, pistils small or developed, petaloid, greenish coloured. Stalks fairly long; flowers upright. Flowering late midseason.

Plant medium to dwarf, partially spreading. Branches relatively slender. Leaves orbicular, medium-sized; leaflets ovate, shallowly lobed, obtuse at the pex; young branchlets, leaves and petioles light green, with high ornamental value. Growth weak, flowers normal. Classic variety.

cv. Cang Jiao

Crown form, sometimes anemone form occurs. Flowers 16cm × 7cm, light red (52-D), soft; the outer petals 3-whorled, thin, usually toothed, suffused with purplish red at the base; the inner petals wrinkled, fairly sparse, some petaloid filaments bearing anthers are present; pistils developed, petaloid, yellowish green. Stalks very short, flowers hidden among leaves. Flowering midseason.

Plant medium to dwarf, partially spreading. Branches stout. Leaves orbicular, large-sized, thick; leaflets ovate, apex acute, upper surface rough, deep green, suffused purple on the margins. Growth vigorous, flowers many. Bred by Zhaolou Peony Garden, Heze in 1968.

cv. Zhao Jun Chu Sai

Crown form. Flowers 14cm × 9cm, faintly pink (65-C; the outer petals 3-4-whorled, flat, toothed irregularly at the apex, red streaks at the base; the inner petals crowded, wrinkled, toothed; stamens completely petaloid; pistils deformed and elongated, stigmas developed, greenish coloured. Stalks slender, slightly pliable, flowers lateral. Flowering midseason.

Plant tall, partially spreading. Branches stout. Leaves orbicular, small-sized, sparse; leaflets ovate, obtuse at the apex, the margins curved upwards. Growth medium, flowers many. Bred by Wangcheng Park, Luoyang in 1969.

Bai Hua Du

Cang Jiao

Zhao Jun Chu Sai

Bai Hua Kui

cv. Hong Zhu Nu

Crown form. Flowers 15cm × 6cm, red (52-A); the outer petals 2-whorled, large, reflexed, suffused with purplish red at the base, the inner petals soft, slightly rugose, fairly regular; pistils small. Stalks fairly long, slightly pliable, flowers lateral. Flowering early midseason.

Plant dwarf, spreading. Branches stout, curved. Leaves long, medium-sized; leaflets long elliptic, acuminate at the apex, pendulous. Growth medium, flowers many. Bred by Team 9, Zhaolou, Heze in 1975.

cv. Zhu Sha Hong

Crown form, sometimes anemone form occurs. Flowers 17cm × 6cm, red faintly pink (52-B), lustrous; the outer petals 4-5-whorled, stiff, suffused with deep purple at the base, inner petals very small, wrinkled, curved, some petaloid filaments bearing anthers are present, a few stamens grow among the petals; pistils small or slightly petaloid. Stalks short, flowers upright or laterally. Flowering midseason.

Plant medium to dwarf, partially spreading. Branches stout and stiff. Leaves long, medium-sized, thick, stiff, sparse; leaflets ovate, acuminate at the apex. Growth medium, flowers many. Bred by Zhaolou Peony Garden, Heze in 1975.

cv. Cai Hui

Crown form. Flowers 16cm × 7cm, light red faintly purple (73-B); the outer petals 2-3-whorled, flat, entire and regular, suffused with purplish red at the base; the inner petals small, wrinkled, crowded, regular, some petaloid filaments bearing anthers are present, a few stamens grow among the petals; pistils small or petaloid. Stalks long, stiff, flowers upright or lateral. Flowering early.

Plant medium to dwarf, partially spreading. Branches fairly slender, stiff. Leaves orbicular, large-sized, crowded; leaflets ovate, thick, soft, slightly pendulous, apex acute. Growth medium; flowers many, neatform. Bred by Zhaolou Peony Garden, Heze in 1973.

cv. Qing Cui Lan

Crown form. Flowers 16cm × 9cm, pink faintly bluish (55-C); the outer petals 2-3-whorled, large, thin, slightly wrinkled, purplish red basal blotches; the inner petals narrowly long, soft, wrinkled, slightly sparse, some petaloid filaments bearing anthers are present, a few stamens grow among the petals; pistils developed, petaloid, yellowish green. Stalks stiff, flowers upright. Flowering early midseason.

Plant medium height, erect. Branches stiff. Leaves orbicular, small-sized, stiff, sparse; leaflets ovate or long-ovate, lobed shallowly, obtuse at the apex, the margins curved upwards. Growth medium; flowers many, flowers neatform; tolerant of adverse conditions. Bred by Heze Flower Garden in 1972.

Hong Zhu Nü

Zhu Sha Hong

Cai Hui

Qing Cui Lan

cv. Shan Hu Tai

Crown form. Flowers 15cm × 10cm, light red (52-C); the outer petals 3-4-whorled, large, thin, dark purple basal blotches; the inner petals rugose, crowded, curved over; pistils small or petaloid. Stalks slender and stiff; flowers upright. Flowering midseason.

Plant dwarf, partially spreading. Branches slender, stiff;. Leaves long, small-sized, crowded; leaflets long ovate, stiff, apex acute. Growth vigorous; flowers many, neatform, each individual long-lasting. Bred by Zhaolou Peony Garden, Heze in 1970.

cv. Xiang Yu

Crown form, sometimes lotus form or anemone form occurs. Flowers 20cm × 7cm, at first light pink, later pure white (155-D); the outer petals 2-whorled, large, flat, suffused with purple at the base; the inner petals regular, crowded, curved over; a few stamens grow among the petals; pistils small or developed, petaloid, greenish coloured. Stalks long, stiff, flowers upright. Flowering midseason.

Plant tall, erect. Branches stout, stiff. Leaves orbicular, large-sized, thick, fairly sparse; leaflets ovate, obtuse at the apex. Growth vigorous; flowers many, several flower forms found on a single plant but unstable; tolerant of diseases and salinity-alkalinity, especially resistant to low temperature during bud stage. Bred by Zhaolou Peony Garden, Heze in 1979.

Xiang Yu

cv. Zhao Fen

Crown form, sometimes lotus form, golden circle form or anemone form occurs. Flowers 18cm × 8cm, pink (38-D); the outer petals 2-3-whorled, large, fairly thin; the inner petals soft, regular, suffused with pinkish red at the base, usually stamens grow among the petals; pistils small or petaloid, sometimes fertile. Stalks fairly stout, long, slightly pliable, flowers lateral. Flowering midseason.

Plant medium height, spreading. Branches fairly pliable, curved. Leaves long, medium-sized, soft, sparse; leaflets long ovate or long elliptic, apex acute, the margins curved upwards. Growth vigorous; flowers many, neatform, high yield of danpi from root for medicine. Classic variety.

Crown Form

Shan Hu Tai

Zhao Fen

cv. Ji Zhao Hong

Crown form. Flowers 15cm × 6cm, red (52-C); the outer petals 2-whorled, soft; the inner petals small, rugose, a few stamens grow among the petals, pistils small or developed, petaloid, greenish coloured. Stalks relatively stiff, flowers lateral. Flowering late midseason.

Ji Zhao Hong

Plant dwarf, partially spreading. Branches relatively slender. Leaves orbicular, medium-sized; leaflets ovate, usually deeply lobed at the apex; upper surface deep green suffused with purple. Growth medium, flowers relatively few. Classic variety.

cv. Ma Ye Hong

Crown form. Flowers 15cm × 5cm, light red (52-C); the outer petals broad and large, curved, suffused with dark purple at the base; the inner petals wrinkled, some petaloid filaments bearing anthers are present. Stalks fairly

Ma Ye Hong

short; flowers lateral. Flowering midseason.

Plant medium height, erect. Branches relatively slender, stiff. Leaves orbicular, small-sized; leaflets broadly ovate, apex acute. Growth vigorous, flowers many. Bred by Team 9, Zhaolou, Heze in 1969.

Yu Guo Tian Qing

cv. Yu Guo Tian Qing

Crown form, sometimes anemone form occurs. Flowers 13cm × 8cm, pink faintly bluish (65-C); the outer petals 2-whorled, large, nearly entire and regular, stiff, dark purple basal blotches; inner petals wrinkled, crowded; pistils small. Stalks short, flowers upright. Flowering midseason.

Plant medium height, partially spreading. Leaves orbicular, medium-sized, soft, relatively crowded; leaflets broadly ovate, acuminate at the apex, pendulous; upper surface deep green, conspicuously suffused with purple. Growth vigorous, flowers relatively few. Classic variety.

cv. Bian Ye Hong

Crown form. Flowers 15cm × 6cm, red (52-C), apex pink; the outer petals 2-3-whorled, large, suffused with red at the base; the inner petals very small, wrinkled, crowded, some petaloid filaments bearing anthers are present; pistils small or developed, petaloid, yellowish green. Stalks short, flowers lateral. Flowering midseason.

Plant dwarf, partially spreading. Branches relatively stout. Leaves orbicular, medium-sized, thick, soft, fairly crowded; leaflets ovate, obtuse at the apex, pendulous, upper surface light or deep green, sometimes these two colours are present on the same leaf. Growth vigorous, flowers relatively few. This cultivar selected from bud mutation of cv.Hu Hong. Classic variety.

cv. Yu Lou Chun Se

Crown form. Flower buds orbicular; flowers 18cm × 10cm, light pink (49-D); the outer petals 2-whorled, large, stiff, suffused with pinkish white at the base; the inner petals long, rugose; pistils developed, petaloid, greenish coloured. Stalks short, slightly pliable, flowers lateral. Flowering late.

Plant medium height, partially spreading. Branches stout. Leaves orbicular, large-sized, crowded; leaflets ovate or broadly-ovate, the margins slightly curved upwards, apex acute. Growth vigorous; flowers many; tolerant

Yu Lou Chun Se

of spring frost, salinity-alkalinity and root diseases. Bred by Zhaolou Peony Garden, Heze in 1995.

cv. Hong Mei Ao Xue

Crown form. Flowers 20cm × 7cm, light red faintly purple (52-D); the outer petals 2-whorled, large, thin, soft, shallowly toothed at the apex, dark purple basal blotches, the inner petals soft, wrinkled, relatively sparse, curved, lighter colour at the apex, a few stamens grow among the petals, smaller. Stalks slender, flowers pendulous. Flowering late midseason.

Plant medium height, spreading. Branches slender, curved. Leaves long,

Bian Ye Hong

Hong Mei Ao Xue

medium-sized, soft; leaflets long ovate or nearly lanceolate, acuminate at the apex, pendulous, the margins curved upwards. Growth vigorous, flowers many, very adaptable. Bred by Team 9, Zhaolou, Heze in 1980.

Man Tian Xing

Qi Zhu Xiang Cui

Bing Ling Zhao Hong Shi

Man Tian Xing

Hong Ling

Jian Rong

cv. Man Tian Xing

Crown form, sometimes anemone form occurs. Flowers 17cm × 9cm, bluish pink (55-D); the outer petals 3-4-whorled, large, stiff, flat, suffused with purple at the base; the inner petals sparse, narrowly long, soft, rugose, some petaloid filaments bearing anthers are present, some stamens grow among the petals; pistils small or developed, petaloid, greenish coloured. Stalks stout, faintly purple, flowers upright. Flowering midseason.

Plant medium height, partially spreading. Branches stout. Leaves orbicular, large-sized, thick, soft; leaflets elliptic, apex acute, the margins curved upwards, upper surface rough, deep green, suffused with pale purple. Growth vigorous; flowers profuse; tolerant of adverse conditions. Bred by Heze Flower Garden in 1979.

cv. Qi Zhu Xiang Cui

Crown form. Flowers 16cm × 8cm, pink (38-C); the outer petals 2-3-whorled, large; stamens completely petaloid; pistils developed, petaloid, yellowish green. Stalks fairly short, flowers upright. Flowering late.

Plant tall, partially spreading. Branches stout. Leaves long, medium-sized, sparse; leaflets long ovate, acuminate at the apex, the margins curved upwards. Growth vigorous, flowers relatively few. Bred by Zhaolou Peony Garden, Heze in 1968.

cv. Hong Ling

Crown form. Flowers 17cm × 7cm, red (52-C), the outer petals 2-whorled, large, entire and regular, with translucent streaks; the inner petals very small, crowded, thin, twisted, some petaloid filaments bearing anthers are present; pistils small. Stalks short, flowers upright. Flowering midseason.

Plant medium height, spreading. Branches fairly stout. Leaves long, medium-sized; petiolules fairly long, slightly pliable, light red; leaflets long ovate, acuminate at the apex, pendulous, the margins ruffled, shallowly lobed, veins sunken conspicuously. Growth relatively weak, flowers many. Bred by Zhaolou Peony Garden, Heze in 1969.

cv. Jian Rong

Crown form. Flowers 16cm × 8cm, light red faintly purple (57-D); the outer petals 3-4-whorled, stiff, suffused with purple at the base; the inner petals crowded, finely toothed, some petaloid filaments bearing anthers are present, pistils small or petaloid. Stalks short, flowers upright. Flowering late.

Plant medium height, partially spreading. Branches stout. Leaves long, medium-sized, thick; leaflets long ovate, acuminate at the apex. Growth vigorous, flower form unusual; very adaptable, resistant to diseases. Bred by Zhaolou Peony Garden, Heze in 1982.

cv. Bing Ling Zhao Hong Shi

Crown form or anemone form. Flowers 16cm × 6cm, light pink (38-D), sparkled; the outer petals 1-2-whorled, thin, apex irregularly toothed, deep purplish red basal blotches; the inner petals twisted, sparse, a few stamens grow among the petals; pistils developed, petaloid, greenish coloured. Stalks short, slightly pliable, flowers lateral. Flowering midseason.

Plant medium height, spreading. Branches slender, slightly curved. Leaves long, large-sized, sparse; petiolules long, leaflets long ovate or obovate lanceolate, obtuse at the apex, pendulous, the margins curved upwards. Growth weak, flowers relatively few. Classic variety.

Tian Zi Guo Se

cv. Tian Zi Guo Se

Crown form. Flowers 20cm × 9cm, light red (50-B); the outer petals 3-whorled, large, entire, toothed at the apex, faintly purplish red basal blotches; the inner petals sparse, soft, wrinkled, some stamens grow among petals, pistils developed, petaloid, greenish coloured. Stalks long, flowers upright. Flowering midseason.

Plant medium height, erect. Branches relatively slender. Leaves long, medium-sized, sparse; leaflets long ovate, apex acute, slightly pendulous, the margins curved upwards. Growth fairly vigorous; flowers profuse, each individual long-lasting. Bred by Zhaolou Peony Garden, Heze in 1976.

cv. Xiao Hu Hong

Crown form. Flower buds orbicular; flowers 15cm × 6cm, red (52-C), bright; the outer petals 3-4-whorled, fairly large, suffused with purplish red at the base; the inner petals small, wrinkled, crowded, a few stamens grow among the petals; pistils small or developed, petaloid, greenish coloured. Stalks fairly stiff, flowers upright. Flowering late midseason.

Plant dwarf, partially spreading. Branches slender. Leaves orbicular, small-sized, thick; leaflets nearly orbicular, with a few and shallow lobes, apex acute. Growth weak, flowers many. Classic variety.

cv. Ruan Yu Wen Xiang

Crown form. Flowers 15cm × 8cm, pink (62-C); the outer petals 2-whorled, large, suffused with red at the base; the inner petals wrinkled, crowded, curved over, some petaloid filaments bearing anthers are present, toothed at the apex; pistils small. Stalk fairly long, pliable; flowers lateral. Flowering midseason.

Plant medium height, spreading. Branches relatively pliable, curved. Leaves orbicular, medium-sized, soft; leaflets long ovate, apex acute, terminal leaflets pendulous. Growth vigorous, flowers many. Bred by Zhaolou Peony Garden, Heze in 1985.

cv. Xiu Wai Hui Zhong

Crown form. Flowers 17cm × 9cm, deep pink (55-D); the outer petals 2-whorled, large, entire and regular, thick, purplish red basal blotches, large; the inner petals very small, rugose, crowded, some petaloid filaments bearing anthers are present; pistils small or short, floral discs purplish red. Stalks fairly long, stiff, flowers upright. Flowering late.

Plant medium height, partially spreading. Branches fairly stiff. Leaves orbicular, medium-sized, fairly sparse; leaflets ovate, apex acute, the margins curved upwards. Growth medium; flowers many, neatform; tolerant of spring frost. Bred by Zhaolou Peony Garden, Heze in 1995.

Xiao Hu Hong

Ruan Yu Wen Xiang

cv. Yin Hong Lou

Crown form. Flowers 16cm × 6cm, light red (52-D), later pink at the apex, lustrous; the outer petals 2-3-whorled, large, soft, suffused with deep purple at the base; the inner petals slightly large, rugose, a few stamens grow among the petals; pistils small or short. Stalks fairly long, slightly pliable; flowers lateral. Flowering early.

Plant dwarf, spreading. Branches slender, slightly pliable. Leaves long, medium-sized; leaflets long ovate, acuminate at the apex, the margins curved upwards. Growth weak, flowers many. Bred by Zhaolou Peony Garden, Heze in 1969.

cv. Yin Hong Ying Yu

Crown form. Flowers 18cm × 8cm, light red (52-D); the outer petals 2-whorled, large, stiff, entire and regular, purplish red basal blotches, the inner petals rugose, crowded, apex pinkish white, a few stamens grow among the petals, pistils small or slightly petaloid. Stalks stout, stiff, flowers lateral. Flowering early.

Plant medium height, partially spreading. Branches relatively stout. Leaves long, large-sized, thick; leaflets long ovate, shallowly lobed, acuminate at the apex, terminal leaflets pendulous. Growth medium, flowers many, neatform, tolerant of strong sunshine. Bred by Zhaolou Peony Garden, Heze in 1968.

Crown Form

Xiu Wai Hui Zhong

Yin Hong Lou

cv. Lan Cui Lou
Crown form. Flowers 16cm × 8cm, pink faintly bluish, apex pinkish white (62-B); the outer petals 2-whorled, large, stiff, suffused with purple at the base, the inner petals regular, crowded, curved; pistils developed, petaloid, yellowish green. Stalks fairly slender, pliable, flowers laterally pendulous. Flowering late.

Plant tall, spreading. Branches stout. Leaves orbicular, large-sized; leaflets ovate, thick, large, soft, pendulous, obtuse at the apex, the margins slightly curved upwards. Growth vigorous; flowers neatform. Bred by Zhaolou Peony Garden, Heze in 1966.

cv. Jin Zhang Fu Rong
Crown form. Flowers 13cm × 6cm, pink (52-D); the outer petals 2-whorled, slightly large, stiff, nearly entire and regular, flat, suffused with red at the base; the inner petals narrowly long, wrinkled; pistils developed, petaloid, greenish coloured. Stalks short, flowers hidden among leaves. Flowering midseason.

Plant medium height, partially spreading. Branches fairly stout. Leaves long, medium-sized, stiff, sparse; leaflets long ovate-lanceolate, the margin curved upwards, acuminate at the apex. Growth vigorous, flowers relatively few. Classic variety.

cv. Shou Zhong Hong
Crown form. Flowers 17cm × 6cm, faintly pink (52-C); the outer petals 2-whorled, broad and large, thin, soft, base red; the inner petals short, twisted and wrinkled, arranged closely, incomplete petaloid stamens grow among the petals; pistils small. Stalks long, stiff, flowers lateral. Flowering midseason.

Plant medium height, spreading. Branches stout, slightly pliable. Leaves long, medium-sized, relatively sparse; leaflets ovate-lanceolate, acuminate at the apex. Growth vigorous, flowers many. Bred by Mr. Zhao Shouzhong at Zhaolou Peony Garden, Heze in 1963.

cv. Zheng Chun
Crown form, sometimes anemone form occurs. Flowers 15cm × 6cm, pinkish red (38-D); the outer petals 3-whorled, stiff, slightly suffused with deep pink at the base; the inner petals wrinkled, relatively few; a few stamens grow among the petals; pistils small or petaloid. Stalks slightly pliable, flowers lateral. Flowering early.

Plant dwarf, spreading. Branches slender, relatively pliable. Leaves long, small-sized, crowded; leaflets long ovate, apex acute, slightly pendulous, upper surface green, fading to yellowish brown at the apex. Growth weak, flowers many. Bred by Zhaolou Peony Garden, Heze in 1983.

cv. Hu Hong
Crown form, sometimes lotus form or anemone form occurs. Flowers 16cm × 7cm, light red (52-C), apex pink; the outer petals 2-3-whorled, large, suffused with deep red at the base; the inner petals soft, wrinkled, crowded, curved over, some petaloid filaments bearing anthers are present; pistils developed, petaloid, greenish coloured. Stalks short, flowers upright or lateral. Flowering late.

Plant medium height, partially spreading. Branches fairly stout. Leaves orbicular, large-sized, thick, crowded; leaflets ovate, obtuse at the apex, pendulous. Growth vigorous; flowers many. Classic variety.

Yin Hong Ying Yu

Lan Cui Lou

Shou Zhong Hong

Jin Zhang Fu Rong

Zheng Chun

Hu Hong

cv. Ping Shi Yan

Crown form. Flowers 12cm × 6cm, deep pinkish red (58-C); the outer petals 3-4-whorled, stiff, flat; the inner petals fairly sparse, wrinkled, some petaloid filaments bearing anthers are present; pistils small. Stalks long, stiff, flowers upright. Flowering midseason.

Plant medium height, partially spreading. Branches relatively stout. Leaves orbicular, medium-sized, thick, stiff, sparse; leaflets nearly orbicular, flat, obtuse at the apex. Growth rather vigorous, flowers many. Classic variety.

cv. Cui Ye Zi

Crown form. Flowers 16cm × 8cm, purple (68-A); the outer petals 3-whorled, entire and regular, stiff, flat, suffused with deep purple at the base; the inner petals regular; pistils small or petaloid. Stalks fairly long, stiff, flowers upright. Flowering midseason.

Plant medium height, erect. Branches fairly stout. Leaves orbicular, medium-sized, thick; stiff, fairly sparse; leaflets ovate, the margins curved upwards, apex obtuse, a few leaflets abnormal. Growth medium, flowers many, young branches and leaves green in early spring. Bred by Team 9, Zhaolou, Heze in 1971.

cv. Zhao Zi

Crown form. Flowers 14cm × 9cm, purple (74-C); outer petals 2-whorled, with numerous teeth, suffused with purplish red at the base; the inner petals stiff, wrinkled, some petaloid filaments bearing anthers are present, a few stamens grow among the petals, pistils small or slightly petaloid. Stalks fairly short, slightly pliable, flowers lateral. Flowering midseason.

Plant medium height, spreading. Branches fairly stout. Leaves long, medium-sized, sparse; leaflets long ovate or elliptic, apex acute, the margins conspicuously curved upwards. Growth fairly vigorous, flowers many, the flower colour gradually faded by sunning. Classic variety developed by Zhao's Sangli Garden during the Qing Dynasty.

Crown Form 167

Ping Shi Yan

Gu Yuan Yi Feng

Zui Xi Shi

Cui Ye Zi

Jiao Hong

Zhao Zi

cv. Gu Yuan Yi Feng
Crown form. Flowers 15cm × 8cm, purple (66-D); the outer petals 3-4-whorled, large, stiff, suffused with red at the base; the inner petals very small, crowded, apex rugose, some petaloid filaments bearing anthers are present, stamens completely petaloid; pistils developed, petaloid, greenish coloured. Stalks short, flowers lateral. Flowering midseason.
Plant dwarf, partially spreading. Branches stout. Leaves long, stiff, crowded; leaflets ovate, apex acute, the margins slightly curved upwards, upper surface green with pink flecks. Growth medium; flowers many; tolerant of strong sunshine. Bred by Gujin Garden, Heze in 1984.

cv. Jiao Hong
Crown form, sometimes anemone form occurs. Flowers 20cm × 7cm, deep pinkish red (52-D); the outer petals 3-4-whorled, stiff, flat, apex pink, suffused with red at the base; the inner petals narrowly long, wrinkled, some petaloid filaments bearing anthers are present, a few stamens grow among the petals; pistils small or slightly petaloid. Stalks relatively stout, stiff, flowers lateral. Flowering midseason.
Plant medium height, spreading. Branches fairly stout. Leaves long, medium-sized, stiff; leaflets ovate or long ovate, apex acute, the margins slightly curved upwards; upper surface deep green, apex green fading to yellowish green. Growth medium, flowers many. Bred by Zhaolou Peony Garden, Heze in 1965.

cv. Zui Xi Shi
Crown form. Flowers 22cm × 10cm, pinkish red (48-C); the outer petals 2-whorled, large, soft, suffused with pale purple at the base, the inner petals narrowly long, wrinkled, fairly large, sparse, apex pinkish white, a few stamens grow among the petals; stigmas enlarged, floral discs purple. Stalks fairly slender, pliable, faintly purple. flowers laterally pendulous. Flowering early midseason.
Plant tall, spreading. Branches pliable, curved. Leaves long, medium-sized, soft, sparse; leaflets long ovate or lanceolate, acuminate at the apex, pendulous. Growth vigorous; flowers many, large; plants swung elegantly in the winds. Bred by Heze Flower Garden in 1968.

Yin Fen Jin Lin

cv. Yin Fen Jin Lin

Crown form. Flowers 15cm × 8cm, pinkish red (62-C); the outer petals 2-whorled, large, nearly entire and regular, blackish purple basal blotches, the inner petals wrinkled, of similar size, regular, crowded; pistils petaloid. Stalks long, slender; flowers pendulous. Flowering late.

Plant dwarf, spreading. Branches slender, curved. Leaves long, medium-sized, soft; leaflets long ovate, shallowly lobed, acuminate at the apex. Growth fairly weak; flowers many, each individual long-lasting. Classic variety.

Zhou Ye Hong

cv. Zhou Ye Hong

Crown form. Flowers 14cm × 5cm, light red (52-C); the outer petals 2-whorled, large, thin, with conspicuous streaks, suffused with red at the base; the inner ones rugose, a few stamens grow among the petals; pistils small. Stalks fairly slender; flowers upright. Flowering midseason.

Plant dwarf, partially spreading. Branches fairly slender. Leaves long, small-sized, fairly sparse; leaflets long ovate, curved downwards, acuminate at the apex. Growth slow; flowers many. Bred by Team 9, Zhaolou, Heze in 1970.

Wei Zi

Chi Lin Xia Guan

Zi Yan Duo Zhu

cv. **Wei Zi**

Crown form. Flowers 12cm × 8cm, purple (74-C); the outer petals 2-whorled, large, stiff, suffused with purple at the base; the inner petals very small, crowded, wrinkled, some petaloid filaments bearing anthers are present; pistils small or absent. Stalks long, stout, stiff, flowers lateral. Flowering late.

Plant dwarf, spreading. Branches slender. Leaves orbicular, small-sized, bi-ternate; leaflets broadly ovate, upper surface light green, suffused with purple on the margins. Plants thin and small, growing slowly; flowers many, neatform. Classic variety.

cv. **Chi Lin Xia Guan**

Crown form or anemone form. Flowers 14cm × 5cm, pale purple faintly bluish (68-B), generally with lighter colour at the apex; the outer petals 2-whorled, ovate, purple basal blotches; the inner petals very small; pistils developed, petaloid, greenish coloured. Stalks stout, straight, pale purplish red. Flowering late.

Plant tall, partially spreading. Branches relatively stout. Leaves orbicular, medium-sized; leaflets ovate, obtuse at the apex, the margins slightly curved upwards. Growth vigorous, flowers many. Bred by Wangcheng Park, Luoyang in 1969.

cv. **Zi Yan Duo Zhu**

Crown form. Flowers 14cm × 7cm, purple, the outer petals 2-3-whorled, large, entire and regular, stiff, suffused with dark purple at the base; the inner petals very small, wrinkled, crowded, some petaloid filaments bearing anthers are present; pistils small or petaloid. Stalks short, flowers upright. Flowering midseason.

Plant dwarf, partially spreading. Branches slender, stiff. Leaves long, small-sized, stiff, sparse; leaflets ovate, apex acute. Growth rather weak, flowers many. Classic variety.

cv. **Wei Hua**

Crown form. Flowers 18cm × 12cm, purple (72-D), with pinkish white tip, slightly lustrous; the outer petals 3-whorled, the inner petals erect, wrinkled, thick, fairly stiff; pistils small. Stalks stout, stiff, flowers upright. Flowering midseason.

Plant medium height, partially spreading. Branches fairly stout. Leaves orbicular, medium-sized; leaflets ovate, obtuse at the apex, suffused with pale purplish red on the margins, with sunken veins, upper surface rough, deep green. Growth vigorous, flowers many. Classic variety.

cv. **Hong Mei Fei Xue**

Crown form. Flowers 18cm × 9cm, pinkish red (49-A); the outer petals 3-whorled, large, soft, thin, suffused with light red at the base; the inner petals narrowly long, irregular, sparse, rugose, apex pinkish white, a few stamens are among the petals; pistils small or slightly petaloid, floral discs purplish red. Stalks slender, stiff, flowers upright or lateral. Flowering midseason.

Plant dwarf, spreading. Branches slender, stiff, slightly curved. Leaves long, small-sized, stiff, crowded; leaflets elliptic or ovate, apex abruptly acute, the margins curved upwards. Growth medium; flowers many, tolerant of adverse conditions, particularly low temperature during bud stage. Bred by Heze Flower Garden in 1969.

cv. **Hong Mei Bao Chun**

Crown form. Flowers 17cm × 8cm, light red (52-C), smooth; the outer petals 2-whorled, large, stiff, suffused with red at the base, apex lighter colour; the inner petals wrinkled, crowded, a few stamens grow among the petals; pistils small or slightly petaloid. Stalks fairly pliable, stout, flowers lateral. Flowering late midseason.

Plant dwarf, partially spreading. Branches stout. Leaves orbicular, large-sized, soft, thick, fairly crowded; leaflets ovate, apex acute. Growth vigorous; flowers normal; tolerant of leaf spot. Bred by Zhaolou Peony Garden, Heze in 1970.

Crown Form 171

Wei Hua

Hong Mei Bao Chun

Hong Mei Fei Xue

Zi Mei You Chun

Zi Mei You Chun

cv. Zi Mei You Chun

Crown form. Flowers 16cm × 7cm, pinkish red faintly purple (68-C); the outer petals 2-whorled, thin, large, suffused with light red at the base; the inner petals soft, sparse, wrinkled, irregular, some stamens grows among the petals; pistils 5, smaller, floral discs pale purple. Stalks stout, long, flowers upright. Flowering midseason.

Plant tall, erect. Branches stout. Leaves long, medium-sized, crowded; petiolules curved; leaflets long ovate, acuminate at the apex, the margins curved upwards. Growth vigorous; flowers many, often 1-2-lateral flowers on a branch blooming 2-3 days later than the terminal ones. Bred by Zhaolou Peony Garden, Heze in 1986.

cv. Zhu Hong Jue Lun

Crown form, sometimes anemone form occurs. Flowers 16cm × 5cm, light red (61-C); the outer petals 2-whorled, large, dark purple basal blotches; the inner petals soft, sparse, arranged irregularly; pistils normal. Stalks slender, stiff, long, flowers upright. Flowering late midseason.

Plant tall, erect. Branches fairly slender, stiff. Leaves orbicular, small-sized, stiff, sparse; leaflets broadly ovate, apex obtuse, suffused with purplish red on the margins. Growth fairly vigorous; flowers many. Classic variety.

cv. Gong Yang Zhuang

Crown form, sometimes anemone or lotus form occurs. Flowers 20cm × 9cm, pinkish red faintly bluish (67-C); the outer petals 1-2-whorled, large, suffused with red at the base; the inner petals relatively narrow, rugose, usually stamens grow among the petals, pistils small. Stalks fairly long; flowers lateral. Flowering early.

Plant medium height, spreading. Branches stout. Leaves long, large-sized, flat; leaflets long ovate, acuminate at the apex. Growth vigorous; flowers many, and several forms. Classic variety.

cv. Sa Jin Zi Yu

Crown form. Flowers 16cm × 9cm, purple (78-D); the outer petals 2-3 whorled, large, stiff, flat, suffused with purple at the base; the inner petals very small, rugose, crowded, some petaloid filaments bearing anthers are present, the upper petals slightly large and wrinkled, nearly all stamens petaloid; pistils developed, petaloid, deep greenish coloured. Stalks short, relatively pliable, pale purple, flowers lateral. Flowering midseason.

Plant medium height, partially spreading. Branches pliable, curved. Leaves long, medium-sized, soft, sparse; leaflets long ovate, apex acute, the margins curved upwards. Growth medium; flowers many. Bred by Heze Flower Garden in 1994.

Crown Form

Gong Yang Zhuang

Zhu Hong Jue Lun

Sa Jin Zi Yu

cv. Mo Lou Cang Jin

Crown form. Flowers 16cm × 7cm, dark purplish red (53-A); the outer petals 3-4-whorled, the margins slightly rugose, lustrous, slightly deep colour at the base, the majority of stamens petaloid, pistils small, occasionally normal. Flowering midseason.

Plant medium height, partially spreading. Branches slender, stiff. Leaves orbicular, medium-sized, fairly thick; leaflets elliptic, apex acute, upper surface green, slightly suffused with pale purple on the margins, veins sunken. Growth medium, flowers normal. Bred by Wangcheng Park, Luoyang in 1995.

cv. Xi Ye Zi

Crown form. Flowers 16cm × 7cm, purple (72-D); the outer petals 2-whorled, broad and large, entire and regular, flat, suffused with purple at the base; the inner ones rugose, regular, a whorl of small petals grow between the outer and the inner petals; pistils small. Stalks stout, long; flowers upright. Flowering late midseason.

Plant medium height, erect. Branches stout. Leaves orbicular, medium-sized, sparse; leaflets nearly orbicular, apex obtuse, the margins curved upwards, slightly suffused with purple, upper surface rough. Growth fairly vigorous, flowers many. Bred by Zhaolou Peony Garden, Heze in 1967.

cv. Wan Shi Sheng Se

Crown form. Flowers 17cm × 8cm, pale purple (64-D); the outer petals 2-whorled, entire and regular, stiff, suffused with purple at the base; the inner petals arranged regularly, apex pinkish purple; a few stamens small; pistils small. Stalks stout, pliable, flowers lateral. Flowering midseason.

Plant medium height, spreading. Branches fairly stout, soft. Leaves long, large-sized, fairly crowded; leaflets long ovate, acuminate at the apex, the margins curved upwards; upper surface deep green, conspicuously suffused with purple. Growth fairly vigorous, flowers many. Bred by Zhaolou Peony Garden, Heze in 1966.

Mo Lou Cang Jin

cv. Ding Xiang Zi

Crown form. Flowers 15cm × 9cm, pinkish purple (74-D); the outer petals 2-whorled, large, suffused with deep purple at the base; the inner petals crowded; pistils small or absent. Stalks long, stiff, flowers lateral. Flowering late.

Plant medium height, spreading. Branches stout. Leaves long, medium-sized, thick; leaflets long elliptic, acuminate at the apex, pendulous, the margins curved upwards. Growth fairly vigorous, flowers many; tolerant of leaf spot. Bred by Zhaolou Peony Garden, Heze in 1968.

cv. Chen Xi

Crown form. Flowers 16cm × 7cm, at first light pinkish purple, later creamy yellow (55-C), usually some petals irregularly purple or pale purple; the outer petals 3-4-whorled, fairly large, entire and regular; the inner petals varied in size, wrinkled, purplish red basal blotches, stamens partly petaloid, pistils developed, petaloid, greenish coloured, floral discs white. Stalks fairly straight, flowers upright. Flowering midseason.

Plant fairly tall, erect. Branches slender, stiff. Leaves long, medium-sized, sparse; leaflets elliptic, acuminate at the apex. Growth medium, flowers many. Bred by Wangcheng Park, Luoyang in 1969.

Crown Form

Xi Ye Zi

cv. Wan Die Yun Feng
Crown form. Flowers 16cm × 8cm, pale purple (63-D); the outer petals 2-whorled, thick, entire, suffused with red at the base; the inner petals crowded, wrinkled, layered, pinkish white with purple tinge when initially flowering, some petaloid filaments bearing anthers are present, a few stamens grow among the petals; pistils 5, floral discs purplish red. Stalks fairly short, slightly pliable, flowers lateral. Flowering midseason.

Plant medium height, spreading. Branches stout, slightly curved. Leaves orbicular, medium-sized, crowded; leaflets ovate or broadly ovate, obtuse at the apex. Growth medium, flowers many. Bred by Zhaolou Peony Garden, Heze in 1973.

cv. Yin Lin Bi Zhu
Crown form. Flowers 15cm × 6cm, pinkish purple (62-B); the outer petals 2-3-whorled, entire and regular, suffused with purple at the base, the inner petals arranged closely, apex lighter colour, with a few anthers, apex curved inwards, scale-like; pistils petaloid. Stalks stiff, flowers lateral. Flowering midseason.

Plant high, erect. Branches stout. Leaves orbicular, large-sized, thick; leaflets ovate, the margins curved and twisted, obtuse at the apex. Growth vigorous, flowers many; tolerant of leaf spot. Bred by Zhaolou Peony Garden, Heze in 1967.

cv. Dan Zao Liu Jin
Crown form. Flowers 16cm × 6cm, purplish red (60-D), sparkling; the outer petals 2-3-whorled, thick, stiff, suffused with dark purple at the base; the inner petals fairly large, sparse, a few stamens grow among the petals; pistils small. Stalks stiff; flowers upright. Flowering midseason.

Plant dwarf, partially spreading. Branches slender. Leaves long, small-sized; leaflets long ovate, apex acute, deeply lobed, the margins conspicuously curved upwards to semi-tube. Growth fairly weak, flowers relatively few. Classic variety.

cv. Fen Zi Guan
Crown form. Flowers 18cm × 12cm, pinkish purple (73-A); the outer petals 1-whorled, large, at first flat, later curved downwards, apex toothed, suffused with purple at the base, the inner petals slightly sparse, slightly wrinkled, regular; stamens petaloid, smaller. Stalks slender, pliable, flowers laterally pendulous. Flowering late.

Plant tall, partially spreading. Branches slender, soft. Leaves long, large-sized, thin, sparse; leaflets long ovate-lanceolate, acuminate at the apex. Growth vigorous, flowers many, large; tolerant of adverse conditions. Bred by Heze Flower Garden in 1979.

cv. Zhong Sheng Zi
Crown form, sometimes anemone form occurs. Flowers 14cm × 6cm, pinkish purple (73-C); the outer petals 2-whorled, large, thin, soft, conspicuously with deep purple streaks, purplish red basal blotches, inner petals soft, rugose, regular, some petaloid filaments bearing anthers are present; smaller or slightly petaloid. Stalks stiff, flowers upright. Flowering midseason.

Plant medium height, partially spreading. Branches slightly stout. Leaves long, small-sized, sparse; leaflets long ovate, apex acute, the margins curved upwards, upper surface rough, deep green, suffused with purple. Growth medium; flowers many, not tolerant of strong sunshine. Classic variety.

176 Chinese Tree Peony

Wan Shi Sheng Se

Ding Xiang Zi

Wan Die Yun Feng

Chen Xi

Crown Form

Yin Lin Bi Zhu

Fen Zi Guan

Zhong Sheng Zi

Dan Zao Liu Jin

San Ying Shi

cv. San Ying Shi

Crown form. Flowers 15cm × 9cm, purplish red (53-B); the outer petals 4-5-whorled, fairly large, stiff, wavily rugose, suffused with deep purplish red at the base; the inner petals very small, crowded, some with large tips, apex pink, a few stamens grow among the petals; pistils normal or short, floral discs pale purplish red. Stalks long, stiff, flowers upright. Flowering early midseason.

Plant medium height, partially spreading. Branches fairly stout. Leaves long, medium-sized, sparse, stiff; leaflets ovate, acuminate at the apex, suffused with pale purple on the margins. Growth vigorous; flowers many, very adaptable. Bred by Zhaolou Peony Garden, Heze in 1970.

Xiang Yang Da Hong

cv. Xiang Yang Da Hong

Crown form. Flowers 19cm × 7cm, red (44-D); the outer petals 2-whorled, broad and large, with translucent streaks, suffused with deep purple at the base; the inner petals wrinkled, a few stamens grow among the petals; pistils small. Stalks long, slightly pliable, flowers lateral. Flowering midseason.

Plant dwarf, spreading. Branches stiff. Leaves long, medium-sized, thick, stiff; leaflets long ovate, apex acute, the margins curved upwards. Growth weak, flowers many. Classic variety, allegedly originated from Xiangyang, Hubei province.

Crown Form 179

Bai Yuan Hong Xia

cv. Bai Yuan Hong Xia

Crown form. Flowers 16cm × 8cm, purplish red (60-B); the outer petals 3-4-whorled, large, stiff, suffused with deep purplish red at the base; the inner petals sparse, long, wrinkled, a few stamens grow among the petals; pistils small or developed, petaloid, greenish coloured. Stalks long, stiff, flowers upright. Flowering midseason.

Plant tall, erect. Branches stiff. Leaves long, medium-sized, stiff, fairly crowded; leaflets long ovate, apex acute. Growth vigorous; flowers many, bright-colour; tolerant of low temperature during bud stage. Bred by Heze Flower Garden in 1972.

Bai Yuan Hong Xia

cv. Xiao Wei Zi

Crown form. Flowers 16cm × 5cm, dark purplish red (53-A); the outer petals 2-3-whorled, thin, soft, suffused with dark purple at the base; the inner petals sparse, rugose, some petaloid filaments bearing anthers are present, a few stamens grow among the petals; pistils small. Stalks fairly short, slightly pliable; flowers hidden among the leaves. Flowering midseason.

Plant medium height, partially spreading. Branches relatively slender, pliable. Leaves long, medium-sized; leaflets long ovate, obtuse at the apex, the margins curved upwards, upper surface green, suffused with pale purplish green. Growth weak, flowers relatively few. Classic variety.

Xiao Wei Zi

Bao Lan Guan

Bao Lan Guan

Mo Kui

Sai Dou Zhu

Teng Hua Zi

Zi Yan Pi Shuang

Si Rong Hong

cv. Bao Lan Guan

Crown form. Flowers 13cm × 11cm, pinkish purple faintly bluish (72-D); the outer petals 2-whorled, large, slightly curved downwards, the inner petals fairly large, erect, suffused with purplish red at the base, with deep purple streaks in petals, a few stamens grown among the petals; pistils normal, stigmas purplish red; floral discs purplish red. Stalks fairly straight, flowers slightly lateral. Flowering midseason.

Plant dwarf, partially spreading. Branches stout. Leaves orbicular, medium-sized, bi-ternate pinnates; leaflets long ovate, acuminate at the apex. Growth medium, flowers many. Bred by Guose Peony Garden, Luoyang in 1992.

cv. Mo Kui

Crown form. Flowers 17cm × 9cm, purple (74-C); the outer petals 2-whorled, large, stiff, dark purple basal blotches; the inner petals wrinkled, some petaloid filaments bearing anthers are present, crowded, curved over; pistils small or petaloid. Stalks fairly pliable; flowers lateral. Flowering midseason.

Plant medium height, spreading. Branches stout, curved. Leaves orbicular, large-sized, large, thick; leaflets broadly ovate, obtuse at the apex. Growth vigorous; flowers many, large, neatform. Classic variety.

cv. Teng Hua Zi

Crown form. Flowers 19cm × 10cm, pinkish purple (63-C); the outer petals 2-3-whorled, broad and large, with conspicuous streaks, with large teeth, suffused with red at the base, stiff; the inner petals long, twisted, crowded, some petaloid filaments bearing anthers are present, a few stamens grow among the petals; pistils small or absent. Stalks stout, stiff, flowers lateral. Flowering midseason.

Plant medium height, partially spreading. Branches stout. Leaves orbicular, medium-sized, thick; leaflets ovate, obtuse at the apex. Growth vigorous, flowers many, neatform; plants elegant, tolerant of spring frost. Bred by Zhaolou Peony Garden, Heze in 1980.

cv. Sai Dou Zhu

Crown form. Flowers 18cm × 8cm, white pinkish purple faintly bluish (68-C); the outer petals 3-whorled, broad and large, the margins toothed, suffused with purple at the base; the inner petals arranged closely, rugose; pistils developed, petaloid, greenish coloured. Stalks relatively pliable, flowers pendulous. Flowering late.

Plant tall, spreading. Branches stout, more curved. Leaves long, medium-sized, sparse; leaflets long ovate, apex acute, pendulous, curved upwards on the margins. Growth vigorous; flowers relatively few; extensive root systems, high yield of danpi from root for medicine. Bred by Zhaolou Peony Garden, Heze in 1967.

cv. Zi Yan Pi Shuang

Crown form. Flowers 14cm × 9cm, faintly pinkish purple (68-B), the margins pinkish white; the outer petals 3-whorled, entire and regular, the inner petals small, arranged closely, pistils developed, petaloid, greenish coloured. Stalks relatively straight, flowers upright. Flowering midseason.

Plant tall, erect. Branches stout. Leaves orbicular, medium-sized, thick; leaflets ovate, with a few and shallow lobes on the margins, suffused with purplish red. Growth vigorous, flowers many. Bred by Wangcheng Park, Luoyang in 1969.

cv. Si Rong Hong

Crown form. Flowers 18cm × 10cm, faintly purplish red (61-C); the outer petals 2-3-whorled, usually toothed, purple at the base; the inner petals crowded, rugose, usually toothed at the apex, the central petals erect, fairly large; pistils small. Stalks stout, flowers upright or lateral. Flowering early midseason.

Plant medium height, erect. Branches stiff. Leaves long, medium-sized, stiff; leaflets ovate, acuminate at the apex, the margins slightly curved upwards. Growth vigorous, flowers many. Bred by Heze Flower Garden in 1995.

cv. Bang Ning Zi

Crown form. Flowers 17cm × 7cm, dark purple, the margins pinkish purple (74-C); the outer petals 2-whorled, large, apex toothed; the inner petals wrinkled, some petaloid filaments bearing anthers are present; pistils developed, petaloid, emerald greenish coloured. Stalks fairly short, flowers upright. Flowering midseason.

Plant dwarf, partially spreading. Branches stout. Leaves orbicular, large-sized, thick, soft; leaflets ovate, apex acute, the margins curved upwards. Growth vigorous. Classic variety bred by Mr. Zhao Bangning in Heze, Shandong province during the Qing Dynasty.

cv. Cui Jiao Rong

Crown form, sometimes single or lotus form occurs. flowers 17cm × 8cm, pinkish white (36-D); the outer petals 3-4-whorled, broad and large, flat, soft; the inner petals crowded, thin, wrinkled, apex slightly white, with a purple streak along the petal, a few stamens grow among the petals, pistils developed, petaloid, greenish coloured. Stalks short, flowers lateral. Flowering early.

Plant medium height, partially spreading. Branches stout. Leaves long, medium-sized; leaflets ovate or long ovate with a few lobes on the margins, slightly suffused with purple, acuminate at the apex. Growth vigorous, flowers many, neatform. Bred by Heze Flower Garden in 1988.

cv. Ou Si Kui

Crown form, sometimes anemone form occurs. Flowers 14cm × 6cm, light pinkish purple faintly bluish (65-D); the outer petals 2-whorled, large, usually toothed, suffused with purple at the base; the inner petals soft, rugose, with purple streaks in petal centre, some petaloid filaments bearing anthers are present; pistils petaloid or smaller. Stalks short, flowers hidden among the leaves. Flowering midseason.

Plant medium to dwarf, partially spreading. Branches stout. Leaves orbicular, medium-sized, thick, soft; leaflets ovate, apex obtuse, the terminal leaflet apex pendulous, upper surface deep green, suffused with purple. Growth vigorous, but flowers relatively few. Classic variety.

Bang Ning Zi

Zi Yao Tai

Ou Si Kui

cv. Zi Xia Dian Jin

Crown form. Flowers 16cm × 7cm, purplish red (63-B); the outer petals 2-3-whorled, usually toothed on the margins; the inner petals wrinkled, crowded, some petaloid filaments bearing anthers are present; pistils petaloid. Stalks relatively short, flowers upright. Flowering midseason.

Plant medium height, partially spreading. Branches relatively stout. Leaves long, medium-sized; leaflets long ovate, acuminate at the apex, the margins curved upwards, terminal leaflets with long petiolules. Growth vigorous, flowers relatively few. Bred by Zhaolou Peony Garden, Heze in 1967.

cv. Zi Yao Tai

Crown form. Flowers 13cm × 8cm, pinkish purple (74-C), the outer petals 3-whorled, large, nearly entire and regular, dark purple basal blotches; the inner petals, crowded, wrinkled, some petaloid filaments bearing anthers are present; pistils small or developed, petaloid, greenish coloured. Stalks slightly long, flowers lateral. Flowering early.

Plant dwarf, spreading. Branches slender, curved. Leaves orbicular, small-sized, stiff, sparse; leaflets ovate, obtuse at the apex, upper surface rough, deep green, suffused with deep purple. Growth slow, flowers many. Bred by Zhaolou Peony Garden, Heze in 1963.

Crown Form 185

Zi Xia Dian Jin

Cui Jiao Rong

cv. Guan Shi Mo Yu

Crown form, sometimes anemone form occurs. Flowers 17cm × 8cm, dark purple (187-A), lustrous; the outer petals 3-4-whorled, stiff, black basal blotches, the inner petals wrinkled, crowded, a few stamens grow among the petals; pistils small or petaloid. Stalks fairly long, stiff, flowers upright. Flowering midseason.

Plant medium to dwarf, erect. Branches fairly stout. Leaves orbicular, medium-sized, thick, soft, relatively crowded; leaflets ovate, apex acute, pendulous. Growth medium; flowers many, growth compact. Bred by Zhaolou Peony Garden, Heze in 1973.

cv. Ying Ying Cang Jiao

Crown form. Flowers 16cm × 9cm, purplish red (67-B); the outer petals 2-3-whorled, large; the inner petals fairly broad, erect, wrinkled; pistils small. Stalks relatively slender, suffused with purple; flowers lateral, hidden among the leaves. Flowering late.

Plant tall, erect. Branches stout. Leaves orbicular, large-sized, bi-ternate pinnate; leaflets ovate, deeply bifid, the margins curved upwards, twisted, deep green above, veins sunken, pilose beneath. Growth vigorous; flowers relatively few. Bred by Wangcheng Park, Luoyang in 1969.

cv. Cang Zhi Hong

Crown form. Flowers 16cm × 7cm, purplish red (53-B), lustrous; the outer petals 2-3-whorled, slightly large, stiff, suffused with dark purple at the base; the inner petals wrinkled, regular, crowded, a few stamens grow among the petals; pistils small or developed, petaloid, greenish coloured. Stalks slender, short, flowers hidden among the leaves. Flowering very early.

Plant dwarf, spreading. Branches slender, stiff. Leaves orbicular, medium-sized, relatively crowded; leaflets ovate, obtuse at the apex, the margins slightly curved upwards, upper surface rough, deep green, suffused with deep purple. Growth vigorous, flowers many; tolerant of diseases. Bred by Zhaolou Peony Garden, Heze in 1963.

Guan Shi Mo Yu

Ying Ying Cang Jiao

Cang Zhi Hong

cv. Zi Hong Cheng Yan

Crown form. Flowers 15cm × 9cm, purplish red, the margins pinkish white (67-B); the outer petals 2-whorled, fairly large, slightly toothed, the inner petals fairly small, crowded, wrinkled, purple basal blotches, most stamens petaloid, pistils normal; floral discs purplish red, stigma red. Stalks straight, flowers upright. Flowering midseason.

Plant medium height, erect. Branches slender, stiff. Leaves orbicular, medium-sized, sparse; petiolules long; leaflets flat, apex obtuse, upper surface fairly rough. Growth medium, flowers many. Bred by Guose Peony Garden, Luoyang in 1992.

cv. Guan Yin Mian

Crown form, often single, lotus or anemone form occurs. Flowers 18cm × 8cm, pinkish white (36-D); the outer petals 3-4-whorled, broad and large, flat, suffused with pink at the base; the inner petals fairly sparse, some normal stamens grow among the petals; pistils normal or slightly petaloid, floral discs deep purplish red. Stalks long, stout, stiff, flowers upright. Flowering very early.

Plant tall, large, erect. Branches stout. Leaves long, medium-sized, sparse, stiff; petiolules raised, stiff; leaflets long ovate, apex acuminate, the margins slightly curved upwards. Growth vigorous, flowers many. Bred by Zhaolou Peony Garden, Heze in 1977.

cv. Zi Yun Xian

Crown form. Flowers 12cm × 6cm, purplish red (73-A); the outer petals 2-whorled, large, nearly entire and regular, suffused with dark purple at the base; the inner petals soft, wrinkled, some stamens grow among petals; pistils small. Stalks pliable, flowers lateral. Flowering very late.

Plant dwarf, spreading. Branches slender, fairly pliable. Leaves long, medium-sized, soft; leaflets broadly lanceolate, apex acuminate, pendulous, the margins curved upwards, suffused with pale purple. Growth vigorous, flowers relatively few. Classic variety.

Zi Hong Cheng Yan

Guan Yin Mian

Zi Yun Xian

cv. Hong Cai Qiu

Crown form. Flowers 15cm × 6cm, light reddish purple (67-C); the outer petals 2-3-whorled, large, fairly thick, suffused with purplish red at the base; the inner petals very small, crowded, some petaloid filaments bearing anthers are present; stamens small; pistils small. Stalks fairly short, slightly pliable, flowers lateral. Flowering midseason.

Plant dwarf, spreading. Branches fairly stout. Leaves orbicular, medium-sized, thick, slightly soft; leaflets elliptic, apex acuminate, pendulous. Growth slow, flowers many. Bred by Zhaolou Peony Garden, Heze in 1963.

cv. Qing Long Xi Tao Hua

Crown form. Flowers 15cm × 7cm, pink (73-C); the outer petals 3-4-whorled, large, flat, suffused with pinkish red at the base, soft; the inner petals crowded, wrinkled, curved inwards, slightly soft, with a purple streak purple along the petal, normal stamens grow among the petals and in the flower centre; pistils developed, petaloid, light greenish coloured. Stalks slender, stiff, flowers lateral. Flowering very early.

Plant medium height, partially spreading. Branches fairly slender. Leaves long, medium-sized; petiolules fairly long, pliable; leaflets ovate or elliptic, slightly suffused with purple on the margins, apex acuminate, slightly pendulous. Growth medium, flowers many. Bred by Heze Flower Garden in 1984.

cv. Fen Lou Chun Se

Crown form, sometimes anemone form occurs. Flowers 19cm × 9cm, pinkish red (62-C); the outer petals 4-5-whorled, large, flat, entire and regular, stiff, with purple streaks in petal centre, suffused with pale purple at the base; the inner petals very small, wrinkled, crowded, some normal stamens grows among petals, some petaloid filaments bearing anthers are present, with more stamens near the flower centre, pistils normal or slightly petaloid, floral discs purplish red. Stalks stout, short, flowers lateral, slightly hidden among the leaves. Flowering midseason.

Plant medium height, partially spreading. Branches stout, stiff. Leaves long, large-sized, thick, relatively crowded; leaflets ovate or long ovate, acuminate at the apex, suffused with pale purple on the margins. Growth vigorous, flowers many. Bred by Zhaolou Peony Garden, Heze in 1995.

Hong Cai Qiu

Qing Long Xi Tao Hua

Fen Lou Chun Se

Tian Xiang Duo Jin

cv. Tian Xiang Duo Jin

Crown form. Flowers 14cm × 6cm, purplish red (61-B); the outer petals 2-whorled, suffused with deep purplish red at the base; the inner petals wrinkled, crowded, a few stamens grow among the petals; pistils small. Stalks slender, stiff, flowers upright. Flowering midseason.

Plant medium height, erect. Branches slender, stiff. Leaves orbicular, small-sized, stiff, thick, sparse; leaflets ovate, apex obtuse, upper surface deep green, suffused with deep purple. Growth medium, flowers many. Classic variety.

Tian Xiang Duo Jin

cv. Yan Long Zi

Crown form. Flowers 16cm × 7cm, dark purple (187-A), lustrous; petals 2-whorled, entire and regular, flat, suffused with dark purple at the base, soft; the inner petals crowded, wrinkled, a few stamens grow among the petals; pistils small or developed, petaloid, yellowish green. Stalks long, straight, flowers upright. Flowering midseason.

Plant dwarf, partially spreading. Branches fairly slender. Winter buds narrowly conical, the top curved, aquiline. Leaves long, medium-sized; leaflets long elliptic, the margins curved upwards, veins conspicuous, deep green above, tomentose beneath. Growth weak; flowers many. Classic variety.

cv. Gong Deng

Crown form. Flowers 16cm × 5cm, purplish red (63-A); the outer petals 2-whorled, large, flat, dark purple basal blotches, the inner petals very small, twisted, crowded, anthers remained at the apex, a few stamens grow among the petals; pistils developed, petaloid, greenish coloured. Stalks slender, slightly pliable, flowers lateral. Flowering late midseason.

Plant tall, spreading. Branches fairly slender, curved. Leaves orbicular, medium-sized, sparse; leaflets ovate, apex obtuse, slightly pendulous. Flowers many. Bred by Zhaolou Peony Garden, Heze in 1967.

cv. Yu Fu Ren

Crown form. Flowers 19cm × 8cm, pinkish white (36-D); the outer petals 3-whorled, broad and large, thin, soft; the inner petals sparse, twisted; petals large and wrinkled near the centre, some stamens grow among the petals; pistils slightly petaloid, green, floral discs light magenta. Stalks slightly pliable, green, flowers lateral. Flowering early midseason.

Plant medium to dwarf, partially spreading. Branches slender, slightly pliable. Leaves long, medium-sized, stiff, slightly sparse; leaflets ovate or long ovate, apex acute. Growth medium, flowers many; the buds, branches and leaves in early spring green, specially admired for leaves in early spring. Bred by Zhaolou Peony Garden, Heze in 1990.

cv. Dong Fang Hong

Crown form. Flowers 16cm × 5cm, purplish red (53-B); the outer petals 3-4-whorled, stiff, flat; the inner petals fairly narrow, apex toothed, some stamens grow among the petals, pistils small. Stalks long, stiff, flowers upright. Flowering midseason.

Plant high, erect. Branches fairly stout. Leaves orbicular, medium-sized, thick, stiff, crowded; leaflets ovate, apex acuminate, the margins curved upwards. Growth vigorous; flowers relatively few; tolerant of diseases. Bred by Zhaolou Peony Garden, Heze in 1967.

cv. Yin Yue

Crown form. Flowers 15cm × 6cm, white (155-D); the outer petals 2-whorled, entire and regular, flat, stiff; the inner petals wrinkled, crowded, a few stamens grow among the petals, some petaloid filaments bearing anthers are present; pistils slightly petaloid, floral discs creamy white. Stalks long, stiff, flowers upright. Flowering late midseason.

Plant tall, erect. Branches slender, stiff. Leaves long, medium-sized, thick, soft, pendulous; leaflets ovate or broadly ovate, apex acute or acuminate, the margins curved upwards, slightly suffused with purple; upper surface green, slightly wrinkled. Growth vigorous; flowers many. Bred by Heze Flower Garden in 1993.

cv. Xia Guang Pu Zhao

Crown form. Flowers 17cm × 7cm, pale purplish red (57-D) with bluish pink; the outer petals 3-4-whorled, large, stiff, flat, suffused with deep red at the base; the inner petals crowded, wrinkled, soft, some petaloid filaments bearing anthers are present, sometimes a few stamens grow among the petals; pistils small or petaloid, floral discs purplish red. Stalks short, slender, flowers lateral. Flowering midseason.

Plant dwarf, partially spreading. Branches stout. Leaves orbicular, large-sized, crowded, thick, stiff; leaflets ovate or broadly ovate; apex acute or obtuse, suffused with pale purple on the margins. Growth medium, flowers many, growth slow, growth compact. Bred by Zhaolou Peony Garden, Heze in 1970.

Yan Long Zi

Gong Deng

Yu Fu Ren

Dong Fang Hong

Yin Yue

Xia Guang Pu Zhao

Zi Bao Guan

Hong Xia Hui

cv. Zi Bao Guan

Crown form. Flowers 13cm × 7cm, purplish red faintly bluish (64-C); the outer petals 3-whorled, large; the inner petals very small, crowded, rugose, pistils developed, petaloid, greenish coloured. Stalks slightly pliable, flowers lateral. Flowering midseason.

Plant medium height, partially spreading. Branches fairly slender. Leaves long, large-sized, sparse; leaflets long elliptic, apex acute, the margins curved upwards. Growth medium, flowers normal. Bred by Wangcheng Park, Luoyang in 1969.

cv. Hong Xia Hui

Crown form. Flowers 17cm × 6cm, purplish red (53-B); the outer petals 2-3-whorled, stiff, entire and regular, suffused with deep purple at the base; the inner petals sparse, wrinkled; pistils developed, petaloid, greenish coloured. Stalks stout, stiff, flowers upright or lateral. Flowering midseason.

Plant medium height, partially spreading. Branches stout. Leaves long, medium-sized, stiff; leaflets ovate or long-ovate, deeply lobed, apex acuminate, the margins curved upwards, upper surface yellowish green, conspicuously suffused with purplish red. Growth vigorous, flowers relatively few. Bred by Zhaolou Peony Garden, Heze in 1965.

cv. Zhuang Yuan Hong

Crown form. Flowers 15cm × 6cm, purplish red, later pinkish white (61-C) at the apex; the outer petals stiff, suffused with purplish red at the base, the inner petals stiff, wrinkled, sparse, some petaloid filaments bearing anthers are present, a few stamens grow among the petals, pistils small or slightly petaloid. Stalks slender, stiff, short, flowers upright or lateral. Flowering nidseason.

Plant medium height, partially spreading. Branches stout. Leaves long, large-sized, thick, crowded; leaflets long ovate, apex obtuse, pendulous, upper surface deep green, suffused with purple. Growth vigorous, flowers relatively few. Classic variety.

cv. Fen He Mian

Crown form. Flowers 14cm × 8cm, pinkish purple (65-A); the outer petals 3-4-whorled, large; the inner petals conspicuously smaller than the outer ones, arranged crowdedly, wrinkled; pistils normal; floral discs purplish red. Stalks long, straight, flowers upright. Flowering late midseason.

Plant tall, erect. Branches stout. Leaves orbicular, medium-sized, thick, crowded; petiolules short; leaflets orbicular, obtuse at the apex; suffused with red on the margins. Growth vigorous, flowers normal. Bred by Wangcheng Park, Luoyang in 1969.

cv. Zi Jin Pan

Crown form. Flowers 14cm × 6cm, deep purplish red (53-A), smooth; the outer petals 2-whorled, large, soft, with irregular teeth, dark purple basal blotches; the inner petals twisted, curved inward, some stamens grow among the petals; pistils normal or petaloid. Stalks stiff, flowers upright. Flowering midseason.

Plant medium height, erect. Branches slender, fairly stiff. Leaves long, small-sized, thin, stiff; leaflets long ovate, the margins curved upwards, apex pendulous, upper surface yellowish green, suffused with purple on the margins. Growth fairly vigorous, flowers many, leaves falling early. Classic variety.

cv. Fen Lou Dian Cui

Crown form. Flowers 15cm × 9cm, pinkish white (36-D); the outer petals 5-whorled, broad and large, obvious, the margins toothed, suffused with pink at the base; the inner petals crowded, petals large and wrinkled at the top; normal stamens grow among the petals, some petaloid filaments bearing anthers are present; pistils developed, petaloid, greenish coloured. Stalks long, straight, flowers upright. Flowering early.

Plant medium height, partially spreading. Branches fairly slender, stiff. Leaves long, medium-sized; leaflets long ovate, terminal ones deeply bifid, lateral ones lanceolate with 2-3-large lobes, apex acuminate. Growth medium, flowers many, neatform. Bred by Heze Flower Garden in 1983.

Zhuang Yuan Hong

Crown Form 193

Fen He Mian

Zi Jin Pan

Fen Lou Dian Cui

cv. Wen Ge

Crown form. Flowers 14cm × 8cm, pale purplish red (57-C); the outer petals 2-whorled, large, suffused with purplish red at the base, the inner petals sparse, soft, twisted and wrinkled, slightly curved, some stamens grow among the petals, pistils developed, petaloid, greenish coloured. Stalks short, flowers upright or lateral. Flowering midseason.

Plant dwarf, partially spreading. Branches slightly curved. Leaves orbicular, medium-sized, fairly crowded; leaflets ovate, acute at the apex. Growth medium, flowers many. Bred by Zhaolou Peony Garden, Heze in 1964.

cv. Lan Hua Kui

Crown form. Flowers 14cm × 10cm, pink (73-C) faintly bluish; the outer petals 3-whorled, entire and regular, flat, suffused with light red at the base, fairly stiff; the inner petals crowded, twisted and wrinkled, toothed at the apex; some petaloid filaments bearing anthers are present; pistils normal; floral discs dark purplish red, stigmas purplish red. Stalks fairly long, slender, stiff, flowers upright. Flowering early.

Plant medium height, partially spreading. Branches fairly stout. Leaves long, medium-sized; leaflets long ovate, apex acute, lateral leaflets 1-3-lobed, the margins slightly curved upwards. Growth vigorous, flowers many, neatform. Bred by Heze Flower Garden in 1979.

Lan Hua Kui

Wen Ge

cv. Zi Hong Zheng Yan

Crown form. Flowers 15cm × 5cm, deep purplish red (60-B), lustrous; the outer petals 2-whorled, ruffled, with dark purple blotches at the base; the inner petals twisted and wrinkled, with large petals near the flower centre, some petaloid filaments bearing anthers are present; pistils developed, petaloid, white. Stalks slender, long, flowers upright. Flowering midseason.

Plant tall, erect. Branches fairly slender. Leaves long, small-sized, sparse; leaflets ovate or broadly ovate, apex acuminate, pendulous, with many lobes in the upper parts of the leaflets, the margins slightly curved upwards. Growth vigorous, flowers many. Petals slow to fall after flowers wither. Bred by Zhaolou Peony Garden, Heze in 1976.

cv. Ling Hua Zheng Chun

Crown form. Flowers 15cm × 7cm, bluish pink (73-C); the outer petals 2-whorled, large, thin, deep purple basal blotches, radial streaked to the apex; the inner petals soft, wrinkled, sparse, some petaloid filaments bearing anthers are present, a few stamens grow among the petals; pistils small or petaloid. Stalks slender and fairly long, flowers lateral. Flowering early.

Plant medium to dwarf, spreading. Branches slender, stiff. Leaves orbicular, small-sized, sparse; leaflets ovate, acuminate at the apex, the margins curved upwards, upper surface deep green, suffused with deep purple. Growth medium; flowers many. Bred by Zhaolou Peony Garden, Heze in 1990.

Crown Form

Zi Hong Zheng Yan

cv. Song Chun Lan
Crown form. Flowers 18cm × 9cm, bluish pink faintly purple (62-C); the outer petals 3-4-whorled, broad and large, suffused with purple at the base, the margins irregular; the inner petals crowded, wrinkled; pistils small. Stalks pale purple, stout, pliable, flowers hidden among the leaves. Flowering late.

Plant medium height, spreading. Branches stout. Leaves long, medium-sized; leaflets broadly lanceolate, thick, apex acute, the margins curved upwards, pilose beneath; terminal leaflets shallowly cut. Growth vigorous; flowers relatively few.

cv. Shou An Hong
Crown form. Flowers 15cm × 10cm, deep purplish red (61-A); the outer petals 2-3-whorled, large, stiff, entire and regular, flat; the inner petals crowded, wrinkled; pistils developed, petaloid, greenish coloured or smaller. Stalks stout, stiff, flowers upright. Flowering late midseason.

Plant tall, erect. Branches stout, stiff. Leaves orbicular, large-sized, thick; leaflets broadly ovate, obtuse at the apex. Growth vigorous, flowers many. This cultivar, is a triploid rare cultivar, deep purplish red roots among the tree peonies. Classic variety.

cv. Ping Hu Qiu Yue
Crown form. Flowers 18cm × 10cm, multi-colour; the outer petals 2-3-whorled, thick, stiff, flat, entire and regular, pink with slightly bluish (65-C), pale purple at the base; the inner petals narrowly long, rugose, crowded, yellowish (159-C), some petaloid filaments bearing anthers are present, pistils 5, floral discs pale purple. Stalks long, stiff; flowers upright. Flowering midseason.

Plant tall, erect. Branches stout. Leaves orbicular, medium-sized, sparse; leaflets ovate or broadly ovate, obtuse at the apex. Growth vigorous; flowers many, tolerant of spring frost, diseases and salinity-alkalinity. Bred by Zhaolou Peony Garden, Heze in 1984.

cv. Zi Rong Jian Cai
Crown form. Flowers 18cm × 8cm, deep purplish red (61-B); the outer petals 4-5-whorled, soft, wavily wrinkled, suffused with dark purple at the base; the inner petals fairly crowded, wrinkled, irregular, usually cut at the apex, a few small stamens grow among the petals; pistils small, 9-10, slightly petaloid, floral discs small and nearly absent, white. Stalks slightly short, rarely suffused with purple, flowers upright or lateral. Flowering midseason.

Plant medium height, partially spreading. Branches stout. Leaves orbicular, medium-sized, fairly stiff, crowded; leaflets ovate or broadly ovate, apex acuminate or acute. Growth vigorous, flowers many, very adaptable. Bred by Heze Flower Garden in 1980.

Ling Hua Zheng Chun

cv. Zhong Sheng Hua

Crown form. Flowers 15cm × 9cm, multi-colour (36-C); the outer petals 2-whorled, large, flat, light pink in the lower part of the petals, suffused with pale purple at the apex, base light yellow, the inner petals sparse, wrinkled, some stamens grows among the petals; stamens small or petaloid; pistils small or petaloid. Stalks pale purple, long, stiff, flowers lateral. Flowering midseason.

Plant dwarf, spreading. Branches slender. Leaves long, small-sized, sparse; leaflets long ovate, apex acute, the margins curved upwards, upper surface deep green, suffused with deep purple. Growth weak, flowers many; slightly tolerant of adverse conditions. Classic variety.

cv. Wang Hong

Crown form. Flowers 14cm × 6cm, deep purplish red (60-B), lustrous; the outer petals 2-whorled, large, stiff; the base and flower of same colour, the inner petals wrinkled, a few stamens grow among the petals; pistils small or petaloid. Stalks short, flowers hidden among the leaves. Flowering midseason.

Plant medium height, erect. Branches slender and stiff. Leaves orbicular, large-sized, thick, soft, relatively crowded; leaflets broadly ovate, obtuse at the apex, pendulous, upper surface rough, deep green, suffused with deep purplish red. Growth medium, flowers relatively few. Classic variety.

cv. Chu Feng Yu

Crown form, usually lotus or anemone form occurs. Flowers 14cm × 5cm, yellowish (158-D), sparkling and bright; the outer petals 2-4-whorled, stiff, flat, with translucent streaks; inner sparse, narrowly long, curved, deeply lobed, some petaloid filaments bearing anthers are present, usually normal stamens grow among the petals; pistils normal, stigmas yellowish green; floral discs white. Stalks fairly long, stiff, flowers upright. Flowering midseason.

Plant medium height, erect. Branches slender, stiff. Leaves orbicular, medium-sized; leaflets ovate or elliptic, slightly suffused with purple. Growth medium, flowers normal. Bred by Zhaolou Peony Garden, Heze in 1986.

Song Chun Lan

Shou An Hong

Ping Hu Qiu Yue

Crown Form

Zi Rong Jian Cai

Zhong Sheng Hua

Wang Hong

Chu Feng Yu

GLOBULAR FORM

All stamens are highly petaloid with shapes and sizes similar to those of normal petals. Pistils petaloid or reduced. The whole flower resembles a Chinese artistic ball.

Jiao Rong San Bian

Globular Form 199

Lü Mu Yin Yu

Lü Mu Yin Yu

Xiu Rong

cv. Jiao Rong San Bian
Globular form. Flowers 16cm × 6cm, at first green, faintly pink (62-D), later pinkish when withering; backs of the outer petals green, the inner petals of similar size, stiff, wrinkled, fairly sparse, some petaloid filaments bearing anthers are present, purplish red basal blotches; pistils small or petaloid. Stalks fairly short, stiff, flowers lateral. Flowering midseason.
Plant medium height, partially spreading. Branches relatively stout. Leaves long, medium-sized, sparse; leaflets long ovate, apex acute, pendulous, upper surface rough, green, suffused with purple. Growth vigorous, flowers relatively few. Classic variety.

cv. Lu Mu Yin Yu
Globular form. Flower buds often opening; flowers 20cm × 12cm, at first light (144-D), later white (155-D); the outer petals 2-whorled, slightly large, stiff, green, suffused with light red at the base; the inner petals stiff, wrinkled, crowded, curved over, stamens completely petaloid; pistils developed, petaloid, greenish. Stalks stout, slightly pliable, flowers lateral. Flowering late.
Plant tall, partially spreading. Branches stout. Leaves long, large-sized, soft, thick; leaflets long ovate, obtuse at the apex, pendulous. Growth vigorous, flowers normal, tolerant of leaf spot. Bred by Heze Flower Garden in 1990.

cv. Xiu Rong
Globular form or crown form. Flowers 16cm × 8cm, at first pinkish red, later pink (38-B); the outer and the inner petals of similar size, crowded, curved, stamens completely petaloid; pistils developed, petaloid, greenish coloured. Stalks stout, stiff, short, flowers lateral. Flowering late.
Plant tall, partially spreading. Branches relatively stout. Leaves long, medium-sized, thick, stiff; leaflets long ovate, apex acute, the margins curvrd upwards. Growth vigorous; flowers relatively few, neatform. Bred by Zhaolou Peony Garden, Heze in 1969.

Yan Luo Fen He

Lü Xiang Qiu

Xue Ying Zhao Xia

cv. Yan Luo Fen He
Globular form. Flowers 16cm × 8cm, pink faintly bluish (65-B); petals soft, rugose, reflexed at the apex, arranged sparsely; pistils developed, petaloid, greenish coloured. Stalks short, flowers upright. Flowering late.

Plant medium height, partially spreading. Branches fairly stout. Leaves orbicular, medium-sized, thick, crowded; leaflets ovate with numerous lobes, apex acute, the margins slightly curved upwards. Growth vigorous; flowers many; tolerant of leaf spot and salinity-alkalinity. Bred by Zhaolou Peony Garden, Heze in 1963.

cv. Lu Xiang Qiu
Globular form, sometimes crown form occurs. Flowers 18cm × 10cm, at first light green, later pink (36-D), the outer and the inner ones of similar size, suffused with pink at the base; the inner petals rugose, crowded, curved over; pistils small or petalois. Stalks stout, fairly long, slightly pliable, flowers lateral. Flowering late.

Plant tall, spreading. Branches stout, curved. Leaves long, large-sized, thick, soft, sparse; petioles 18cm long, stout, oblique; leaflets long ovate-lanceolate, acuminate at the apex, upper surface deep green. Growth extremely vigorous; flowers many, large, neatform, tolerant of diseases and salinity-alkalinity. Bred by Zhaolou Peony Garden, Heze in 1975.

Lan Tian Piao Xiang

Fen Yu Qiu

cv. Xue Ying Zhao Xia

Globular form. Flowers 16cm × 12cm, at first pink, later pinkish white (56-D); petals fairly thin, suffused with pink at the base, irregular, fairly crowded, curved over; stamens petaloid; pistils petaloid. Stalks long, pliable, flowers pendulous. Flowering late.

Plant tall, spreading. Branches fairly stout, curved. Leaves long, large-sized, fairly crowded; leaflets ovate or long ovate, deeply lobed, acuminate at the apex, the margins slightly curved upwards, upper surface smooth, deep green. Growth vigorous, flowers relatively few. Bred by Zhaolou Peony Garden, Heze in 1973.

cv. Lan Tian Piao Xiang

Globular form. Flowers 14cm × 11cm, bluish pink (69-A); the outer petals 3-4-whorled, entire and regular, slightly curved downwards or flat, inner and the outer petals of similar size, erect, sparse, stamens completely petaloid; pistils developed, petaloid, deep greenish coloured. Stalks fairly straight; flowers upright. Flowering late midseason.

Plant medium height, erect. Branches stout. Leaves orbicular, medium-sized, thick; leaflets broadly ovate, with a few and shallow lobes, apex obtuse, suffused with purple, the margins slightly curved downwards. Growth vigorous, flowers relatively few. Bred by Wangcheng Park, Luoyang in 1969.

cv. Fen Yu Qiu

Globular form. Flowers 13cm × 8cm, pink (73-C) with bluish; the outer petals 4-whorled, stiff, flat, bluish purple basal blotches ; the inner petals broad and large, the inner and the outer ones of similar size, layered, crowded, wrinkled, with radical streaks purple, normal or small stamens grow among petals; pistils small or petaloid, floral discs purplish red. Stalks fairly slender, short, flowers upright. Flowering early.

Plant dwarf, partially spreading. Branches slender, stiff. Leaves orbicular, medium-sized; petioles 12cm long; leaflets ovate or long ovate, apex acuminate, the margins slightly curved upwards, suffused with purple; upper surface rough, deep green, suffused with purple. Growth medium, flowers many, neatform, tolerant of strong sunshine. Bred by Zhaolou Peony Garden, Heze in 1991.

Crown Proliferate-Flower Form

Two or more individual flowers of Crown Proliferate Form overlapping.

Jin Xiu Jiu Du

Jin Xiu Jiu Du

cv. Jin Xiu Jiu Du

Crown proliferate form. Flowers 16cm × 9cm, red(61-C). Outer petals 3-whorled, broad and large, broadly ovate, suffused with purplish red at the base, most stamens petaloid; pistils developed, petaloid, greenish coloured in the lower flower. Petals very small, rugose; stamens normal; pistils normal in the upper flower. Pedicels stout, straight, flowers upright. Flowering midseason.

Plant medium height, partially spreading. Branches stout. Leaves orbicular, small-sized, stiff; leaflets elliptic, acute at the apex, the margins slightly curved upwards, veins sunken. Growth vigorous, flowers many. Bred by Wangcheng Park, Luoyang in 1969.

cv. Yu Lou Dian Cui

Crown proliferate form. Flowers 17cm × 13cm, white(155-D). Petals multi-whorled, soft, with irregular teeth at the apex, suffused with pink at the base, stamens petaloid or partially infertile; pistils developed, petaloid, greenish coloured in the lower flower. Petals relatively few, large, erect,

Yu Lou Dian Cui

Xian Tao

stamens petaloid or partly infertile; pistils petaloid or partly infertile in the upper flower. Stalks long, fairly pliable, flowers lateral or laterally pendulous. Flowering late.

Plant tall, spreading. Branches stout, pliable, curved. Leaves long, large-sized, sparse, thick; leaflets ovate or long ovate, deep lobed, acuminate at the apex, the margins curved upwards, upper surface deep green. Growth vigorous, flowers relatively few; high yield of danpi from roots for medicine; tolerant of leaf spot; leaves falling late. Bred by Zhaolou Peony Garden, Heze in 1966.

cv. Xian Tao

Crown proliferate form. Flowers 19cm × 9cm, red (52-A)with pink. Petals multi-whorled, soft, thin, purplish red basal blotches, all stamens petaloid; pistils developed into colourful, striped petals in the lower flower. Petals relatively few, large, long, very small near the flower centre, all stamens small or petaloid, pistils small, nearly absent in the upper flower. Stalks fairly long, slender, pliable, flowers lateral. Flowering late midseason.

Plant tall, partially spreading. Branches stout. Leaves long, medium-sized, crowded, thick, pendulous; leaflets ovate or broadly ovate, apex acuminate or acute, the margins wavy and curved upwards. Growth extremely vigorous, flowers relatively few,

Xiu Tao Hua

tolerant of leaf spot. Bred by Heze Flower Garden in 1985.

cv. Xiu Tao Hua

Crown proliferate form. Flowers 18cm × 6cm, pinkish red faintly bluish (73-B), later lighter colour at the apex. The outer petals 3-4-whorled, thick, flat, with conspicuous streaks, the inner petals small, wrinkled, normal or some petaloid filaments bearing anthers are present, pistils small or short in the lower flower. Petals fairly large, stamens a few, pistils small in the upper flower. Stalks fairly long, flowers lateral. Flowering midseason.

Plant medium height, partially spreading. Branches fairly stout. Leaves long, medium-sized, fairly crowded; leaflets long ovate, acuminate at the apex, the margins slightly curved upwards; upper surface green, suffused with purple. Growth vigorous; flowers many; tolerant of leaf spot, leaves falling late. Bred by Zhaolou Peony Garden, Heze in 1969.

cv. Lan Yue

Crown proliferate form. Flowers 18cm × 4cm, pink faintly bluish(73-C). Out petals 2-whorled, nearly entire and regular, purplish red basal blotches, the inner petals soft, slightly rugose; a few stamens normal; pistils developed, petaloid, yellowish green in the lower flower. Petals relatively few, wrinkled; stamens small; pistils small or slightly petaloid in the upper flower. Stalks slightly long and pliable, flowers lateral. Flowering midseason.

Plant tall, erect. Branches fairly slender. Leaves orbicular, medium-sized, fairly sparse; leaflets ovate, obtuse at the apex. Growth vigorous; flowers many. Bred by Zhaolou Peony Garden, Heze in 1990.

cv. Guan Qun Fang

Crown proliferate form. Flowers 18cm × 11cm, deep purplish red (59-B). The outer petals 3-4-whorled, soft, slightly flat, the inner petals very small, wrinkled, crowded, many anthers remained at the apex, some petaloid filaments bearing anthers are present; pistils developed, petaloid mixed red and green in the lower flower. Petals relatively few, wrinkled, erect, pistils and stamens small. Stalks slightly long, stiff, flowers upright or lateral. Flowering midseason.

Plant medium height, partially spreading. Branches stiff. Leaves long, small-sized, soft; leaflets long elliptic or broadly lanceolate, terminal leaflets pendulous, acuminate at the apex, slightly suffused with purple on the margins, upper surface deep green. Growth medium; flowers many, tolerant of low temperature during bud stage. Bred by Heze Flower Garden in 1971.

cv. Chi Long Huan Cai

Crown proliferate form. Flowers 16cm × 8cm, purple (74-C) with pink; petals of the similar size, stiff, crowded, layered, purplish red basal blotches, stamens completely petaloid, pistils developed, petaloid, greenish coloured in the lower flower. Petals relatively few; stamens small or absent; pistils small or absent in the upper flower. Stalks long, flowers lateral. Flowering late.

Plant medium height, spreading. Branches stout, curved. Leaves orbicular, small-sized, stiff, sparse; leaflets nearly orbicular, obtuse at the apex, the margins curved upwards, suffused with purple. Growth fairly vigorous; flowers many. Classic variety.

cv. Tao Hong Xian Mei

Crown proliferate form. Flowers 13cm × 6cm, pinkish red (52-D). The outer petals 2-3-whorled, slightly large, dark purple basal blotches, the inner petals wrinkled, crowded; pistils developed, petaloid, greenish coloured in the lower flower. Petals large, a few, erect; stamens completely petaloid or small; pistils completely petaloid or small in the upper flower. Stalks slender, short. Flowers lateral. Flowering early.

Plant dwarf, partially spreading. Branches slender, pliable. Leaves long, small-sized, crowded; leaflets long ovate or elliptic, apex obtuse, upper surface yellowish green, the margins twisted, suffused with deep purple. Growth weak; flowering rate slightly low, flowers small. Classic variety.

cv. Sheng Dan Lu

Crown proliferate form. Flowers 13cm × 8cm, pinkish red with purple (68-A); outer and the inner petals of similar size; stamens completely petaloid; pistils completely petaloid. The outer petals 2-whorled, flat, the inner petals wrinkled, crowded, stamens completely petaloid, pistils developed, petaloid, greenish coloured in the lower flower. Petals relatively few, slightly large, rugose in the upper flower. Stalks short, flowers lateral. Flowering late.

Plant tall, spreading. Branches stout, stiff. Leaves long, large-sized, large, crowded; leaflets long ovate, acuminate at the apex. Growth extremely vigorous, but flowers relatively few. Classic variety.

Chi Long Huan Cai

Lan Yue

Tao Hong Xian Mei

Guan Qun Fang

Sheng Dan Lu

cv. Ceng Lin Jin Ran

Crown proliferate form. Flowers 16cm × 6cm, purplish red (53-B). The outer petals 3-whorled, thick, suffused with dark purple at the base, the inner petals small, twisted and rugose, pistils developed, petaloid, greenish coloured in the lower flower. Petals fairly large, a few, erect; stamens small or absent; pistils small or absent in the upper flower. Stalks long, stiff, flowers upright. Flowering midseason.

Plant tall, erect. Branches fairly slender. Leaves long, medium-sized; leaflets long ovate, acuminate at the apex. Growth vigorous, flowers many. Bred by Zhaolou Peony Garden, Heze in 1967.

cv. Fen Lou Tai

Crown proliferate form. Flowers 18cm × 10cm, pink (62-D). Petals 2-whorled, large, flat, suffused with pinkish red at the base, the inner petals wrinkled, regular, crowded; stamens completely petaloid; pistils developed, petaloid, greenish coloured in the lower flower. Petals relatively few, thin, stamens and pistils small and absent or petaloid in the upper flower. Stalks stout, pliable, flowers laterally pendulous. Flowering late.

Plant tall, partially spreading. Branches stout. Leaves long, large-sized, thick; leaflets long ovate with numerous lobes, acuminate at the apex. Growth vigorous; flowers many, neatform; tolerant of leaf spot. Bred by Heze Flower Garden in 1982.

cv. Jia Ge Jin Zi

Crown proliferate form. Flowers 15cm × 9cm, purple (73-A). The outer petals 2-whorled, large, flat, suffused with deep purple at the base, the inner petals stiff, wrinkled, crowded; stamens completely petaloid; pistils developed, petaloid, purplish in the lower flower. Petals relatively few, slightly large; stamens petaloid or small and absent; pistils petaloid or small and absent in the upper flower. Stalks long, slightly pliable, flowers lateral. Flowering late.

Plant medium height, erect. Branches fairly stout. Leaves long, medium-sized, thick, stiff; leaflets long ovate or long elliptic, acuminate at the apex, the margins curved upwards. Growth medium; flowers relatively few, neatform. Classic variety.

cv. Zi Rong Qiu

Crown proliferate form. Flowers 18cm × 11cm, dark purplish red (64-C). Petals 3-whorled, large, stiff, flat, the inner petals fairly small, wrinkled; some petaloid filaments bearing anthers are present; pistils developed, petaloid, greenish coloured in the lower flower. Petals relatively few, rugose, nearly all stamens petaloid; pistils developed, petaloid, greenish coloured in the upper flower. Stalks stout, long, flowers lateral. Flowering late midseason.

Plant tall, partially spreading. Branches stout. Leaves orbicular, medium-sized, stiff, thick, fairly crowded; leaflets broadly ovate, apex acute, the margins slightly curved upwards. Growth vigorous, flowers many, tolerant of adverse conditions. Bred by Gujin Garden, Heze in 1982.

cv. Jin Xiu Qiu

Crown proliferate form. Flowers 15cm × 12cm, deep purplish red (61-A). The outer petals 2-3-whorled, large, stiff, suffused with deep purple at the base, the inner petals wrinkled, thick, some petaloid filaments bearing anthers are present, pistils developed, petaloid, purplish red in the lower flower. Petals fairly large, erect, pistils small or petaloid in the upper flower. Stalks long, stiff, flowers upright. Flowering midseason.

Plant medium height, erect. Branches stout. Leaves long, small-sized, stiff, fairly crowded; leaflets long ovate, acuminate at the apex. Growth vigorous; flowers many, neatform. Bred by Team 9, Zhaolou, Heze in 1982.

Ceng Lin Jin Ran

Fen Lou Tai

Zi Rong Qiu

Jia Ge Jin Zi

Jin Xiu Qiu

cv. Hong Yu Lou

Crown proliferate form. Flowers 16cm × 9cm, light red (52-C). The outer petals 6-whorled, broad and large, entire and regular; the inner petals long, rugose, some petaloid filaments bearing anthers are present; pistils developed, petaloid, greenish coloured in the lower flower. Petals relatively few, slightly large; stamens short; pistils short in the upper flower. Stalks fairly short, flowers lateral. Flowering midseason.

Plant tall, spreading. Branches stout, slightly pliable. Leaves long, medium-sized, slightly soft; leaflets ovate or broadly ovate, obtuse at the apex, pendulous, the margins wavy; terminal leaflets deeply trilobed with long petiolules. Growth vigorous, flowers many; tolerant of spring frost and diseases. Bred by Zhaolou Peony Garden, Heze in 1983.

Hong Yu Lou

cv. Song Chun

Crown proliferate form. Flowers 17cm × 6cm, purplish red (61-C); Petals 3-4-whorled, entire and regular, flat, suffused with deep purple at the base, nearly all stamens petaloid, some petaloid filaments bearing anthers are present; pistils occasionally developed, petaloid, greenish coloured in the lower flower. Petals relatively few, wrinkled; stamens infertile or petaloid; pistils small or petaloid in the upper flower. Stalks long, stiff, flowers upright. Flowering late midseason.

Plant medium height, erect. Branches fairly stout, straight, stiff. Leaves long, large-sized, thick, fairly crowded; leaflets long ovate or broadly ovate-lanceolate, apex acuminate. Growing vigorous. Bred by Zhaolou Peony Garden, Heze in 1964.

cv. Ying Luo Bao Zhu

Crown proliferate form. Flowers 15cm × 8cm, light red (58-C). The outer petals 2-3-whorled, stiff, suffused with red at the base; the inner petals wrinkled, thick, with numerous teeth at the apex; pistils developed, petaloid, mixed red and green in the lower flower. Petals slightly large, a few, erect, irregular in the upper flower. Stalks slender, short, flowers upright. Flowering midseason.

Plant dwarf, partially spreading. Branches fairly slender. Leaves long, small-sized, crowded, stiff; leaflets ovate elliptic or long elliptic, apex acuminate. Growth medium, flowers relatively few. Classic variety.

cv. Zhong Sheng Hong

Crown proliferate form. Flowers 14cm × 6cm, red (43-A). Petals multi-whorled, stiff, arranged closely and irregularly, suffused with deep purple at the base, most stamens petaloid; pistils petaloid in the lower flower. Petals relatively few, wrinkled, slightly large with numerous lobes at the apex; stamens small; pistils small in the upper flower. Stalks short. Flowers lateral. Flowering late midseason.

Plant dwarf, partially spreading. Branches fairly stout. Leaves orbicular, medium-sized, thick, crowded; leaflets ovate, apex acute, the margins curved upwards. Growth medium; flowers relatively few. Classic variety.

cv. Jin Hong

Crown proliferate form. Flowers 20cm × 9cm, pale pinkish red (63-B). Petals 4-5-whorled, large, flat, purple basal blotches, most stamens petaloid; pistils developed, petaloid, greenish coloured in the lower flower. Petals wrinkled, slightly large, a few; stamens small; pistils small in the upper flower. Stalks stiff, flowers upright or lateral. Flowering late midseason.

Plant tall, partially spreading. Branches stiff. Leaves orbicular, medium-sized, stiff, thin; leaflets elliptic or lanceolate, acuminate at the apex. Growth vigorous; flowers many; tolerant of leaf spot and low temperature during bud stage. Bred by Heze Flower Garden in 1982.

Song Chun

Ying Luo Bao Zhu

Crown Proliferate-Flower Form

Zhong Sheng Hong

Jin Hong

REFERENCE

1. 仲殊. 越中牡丹花品. 中国农学书录. 986
2. 胡元质. 牡丹记. 古今图书集成. 1011
3. 欧阳修. 洛阳牡丹记. 古今图书集成. 1034
4. 李英. 吴中花品. 中国农学书录. 1045
5. 周师厚. 鄞江周氏洛阳牡丹记. 古今图书集成. 1082
6. 张峋. 洛阳花谱. 古今图书集成. 1086~1094
7. 张邦基. 陈州牡丹记. 古今图书集成. 1111~1117
8. 陆游. 天彭牡丹谱. 古今图书集成. 1178
9. 薛凤翔. 亳州牡丹史. 古今图书集成. 1617
10. 王象晋. 群芳谱. 古今图书集成. 1621
11. 苏毓眉. 曹南牡丹谱, 古今图书集成. 1669
12. 陈淏子. 花镜, 中华书局, 1688
13. 汪灏等. 广群芳谱, 商务印书馆, 1708
14. 吴其濬. 植物名实图考长编. 商务印书馆, 1828
15. 赵世学. 新编桑篱园牡丹谱. 1912
16. 黄岳渊. 花经. 新纪元出版社, 1949
17. 方文培. 中国芍药属的研究. 植物分类学报, 1958, 7 (4): 297~313
18. 潘开玉. 芍药亚科芍药属. 中国植物志. 北京: 科学出版社, 1976, 27: 37~59
19. 国家环保局, 中国科学院植物研究所. 中国珍稀濒危保护植物名录. 北京: 科学出版社, 1978
20. 王世端等. 洛阳牡丹. 北京: 中国旅游出版社, 1981
21. 王莲英等. 牡丹及其栽培品系种的染色体组成. 北京林学院学报, 1983
22. 李保兴. 曹州牡丹史话. 济南: 山东友谊书社, 1987
23. 刘淑敏, 王莲英等. 牡丹. 北京: 中国建筑工业出版社, 1987
24. 张益民, 王进涛等. 河南紫斑牡丹的生境及分布规律的研究. 豫西农专学报, 1988, (1): 4~13
25. 舒迎澜. 清·计楠的牡丹栽培技术. 大众花卉, 1988, (2): 10
26. 汤忠皓. 牡丹花考. 中国园林, 1989, 5 (2): 20~25
27. 喻衡. 牡丹花. 上海: 上海科学技术出版社, 1989
28. 李嘉珏. 临夏牡丹. 北京: 北京科学技术出版社, 1989
29. 狄维忠, 于兆英等. 陕西省第一批珍稀濒危保护植物. 西北大学出版社, 1989
30. 傅立国主编. 中国珍稀濒危植物. 上海: 上海教育出版社, 1989
31. 钱敏之, 张炳坤等. 农架野生紫斑牡丹引种调查研究. 武汉植物学研究, 1991, 9 (4): 372~377
32. 黄砥龙. 牡丹花露地引种栽培的回顾与展望. 东北园林, 1991, (1-2): 48~54
33. 洪涛, 张家勋, 李嘉珏等. 中国野生牡丹研究(一): 芍药属牡丹组新分类群. 植物研究, 1992, 12, (3): 223~234
34. 洪涛, 齐安·鲁普·奥斯蒂. 中国野生牡丹研究(二): 芍药属牡丹组新分类群. 植物研究, 1994, 14 (3): 237~240
35. 裴颜龙, 洪德元. 卵叶牡丹——芍药属一新种. 植物分类学报, 1995, 33 (1): 91~93
36. 温新月, 李保光主编. 国花大典. 济南: 齐鲁书社, 1996
37. Wister, J C. The Peonies. Washington: American Horticultural Society, 1962
38. Haworth-Booth, M. The Moutan or Tree Peony. London: Constable Publishers, 1963
39. Bean, W J. Trees and Shrubs Harding in the British Isles (8th). London: John Murray, 1980
40. Haws, S G. A Problem of Peonies. Garden, 1986, 117 (7): 326~328
41. Haus, S G and Lauener, L A. A Review of the Infraspecific Taxa of *Paeonia suffruticosa* Andr. Edinb. J Bot, 1990, 47 (3): 273~281
42. Harding, A. The Peony. Timber Press, Inc. Oregon, 1993
43. Gessenich, G M. The Best of 75 Years. Amercian Peony Society, MN. 1993

Index of Mudan Cultivar Names

A

Ao Yang   Majestic Sun   **94, 95**

B

Ba Bao Xiang   Mounted Gems   **68**
Bai He Wo Xue   Crane Standing in Snow   **151**
Bai Hua Kui   Flower Queen   **156**
Bai Hua Cong Xiao   Smiling in Clustered Flowers   **92, 93**
Bai Hua Du   The Envy of the Multitude   **156**
Bai Lian Xiang   Fragrance of White Lily   **141, 142**
Bai Tian E   White Swan   **96**
Bai Yu Lan   Yulan Magnolia   **52**
Bai Yu   White Jade   **143**
Bai Yuan Hong Xia   Rosy Sunshine over Gardens   **179**
Bang Ning Zi   Bangning Purple   **184**
Bao Gong Mian   Official Baozheng's Dark Purple Face   **80, 81**
Bao Lan Guan   Sapphire Blue Crown   **182, 183**
Bi Bo Xia Ying   Sunrays Reflecting on Sea Ripples   **76**
Bi Hai Fo Ge   Buddhist Temple at Sea   **148**
Bi Yue Xiu Hua   Outshining the Moon and Other Flowers   **82**
Bian Ye Hong   Leaves Turning Red   **161**
Bing Hu Xian Yu   Jade Tribute in Icy Pot   **142, 143**
Bing Ling Zhao Hong Shi   Ruby Wrapped in Ice   **162, 163**
Bing Ling Zi   Icy Violet   **151**
Bing Zhao Lan Yu   Ice-Cased Blue Jade   **145**

C

Cai Die   Colourful Butterfly   **58**
Cai Hui   Brightly-coloured Painting   **157**
Can Xue   Melting Snow   **147**
Cang Jiao   A Beauty in Seclusion   **156**
Cang Zhi Hong   Red Flower Hidden Among Twigs   **186**
Cao Zhou Hong   Caozhou Red   **114**
Ceng Lin Jin Ran   Garnet Tinted Forests   **205**
Ceng Zhong Xiao   Smiling in Fields of Flowers   **94, 95**
Chao Yi   Court Dress   **126**
Chen Hui   Morning Lights   **90**
Chen Xi   Morning Twilight   **174, 176**
Chi Ling Xia Guan   Shiny Decoration on Lilac Crown   **170**
Chi Long Huan Cai   Burning Dragon with Flashing Brilliance   **204**
Chi Tang Xiao Yue   Lingering Moon over Pond   **54**
Chi Yang   Broiling Sun   **54**
Chong Lou Dian Cui   Towers in Stippled Green   **128**
Chu E Huang   Yellow Gosling Crest   **132**
Chu Feng Yu   Feather of Young Phoenix   **196, 197**
Chun Gui Hua Wu   Spring Returns to the Beautiful House   **126**
Chun Hong Jiao Yan   Tender and Glamorous Spring Pink   **80, 81**
Chun Hong Zheng Yan   Spring Red Contending with Fascination   **71**
Chun Lian   Spring Lotus   **64**
Chun Se Man Yuan   Garden with Spring Beauties   **99**
Cong Zhong Xiao   Beaming in Clusters   **82, 83**

Cui Jiao Rong   Tender Blossom with Delicate Leaves   **184, 185**
Cui Mu   Verdant Curtain   **153**
Cui Ye Zi   Purple above Green Leaves   **166, 167**

D

Da Ban Hong   Huge Purple Petals   **62**
Da Hong Duo Jin   Royal Scarlet Winning Champion   **107**
Da Hong Yi Pin   Top Rank Red Court Dress   **96, 97**
Da Hu Die   Giant Butterfly   **79**
Da Jin Fen   Great Golden Pink Petals   **62, 63**
Da Tao Hong   Huge Peach Blossom Red   **74, 75**
Da Zong Zi   Huge Red-brown Petals   **105**
Dan Lu Yan   Flames in the Furnace   **107**
Dan Ou Si   Light Lotus-root Threads   **132**
Dan Yang   Crimson Sun   **76**
Dan Zao Liu Jin   Gold Running on Plum   **175, 177**
Di Yuan Chun   Spring in the Imperial Garden
Ding Xiang Zi   Lilac   **174, 176**
Dong Fang Hong   Red in the East   **190**
Dong Fang Jin   Oriental Brocade   **123, 125**
Dong Hai Lang Hua   Sea Spray in the East   **86, 87**
Dou Kou Nian Hua   A Budding Beauty   **95**
Dou Lü   Green of Beans   **146, 147**
Dou Zhu   Pleasing Pearls   **152**
Du Juan Hong   Azalea   **112**

E

E Mei Xian Zi   Fairy Maiden in Mount Emei   **153**
Er Qiao   Senior and Junior Sisters Qiao   **104, 105**

F

Fei Yan Hong Zhuang   Flying Swallow Lady in Red   **118**
Fei Yan Zi   Flying Swallow Purple   **56**
Fen E Jiao   An Enervated Palace Lady   **62, 64**
Fen He Mian   Face of Pink lily   **191, 193**
Fen Lan Lou   Pinkish Blue Mansion   **151**
Fen Lou Chun Se   Spring in Powdery Pavilion   **188**
Fen Lou Dian Cui   Spotted Green in Powdery Pavilion   **191, 193**
Fen Lou Tai   Powdery Pavilion   **205**
Fen Mian Tao Hua   Peach-blossomed Face   **136**
Fen Pan Jin Qiu   Glorious Balls in Agate Tray   **148**
Fen Pan Tuo Gui   Osmanthus in Hermosa Plate   **134, 135**
Fen Qing Shan   Rouge-cyan Mountains   **142, 143**
Fen Yu Qiu   Powdery Ball   **201**
Fen Zi Guan   Pink Opera Crown   **175, 177**
Fen Zi Han Jin   Livid Pink Containing Gold   **66, 67**
Fen Zhong Guan   VIP Rose   **150**
Feng Hua Zi   Light Puniceus Lush Flowers   **96**
Feng Ye Hong   Red Maple Leaves   **101**
Fu Gui Man Tang   Full Presentation of Prosperity   **128, 129**

G

Ge Jin Zi   Spirt Gejin in Purple   84
Gong Deng   A Palace Lantern   190
Gong E Jiao Zhuang   Palace Maid in Fine Dress   75
Gong Yang Zhuang   Palatial Adornments   172, 173
Gu Ban Tong Chun   Old Friends Sharing Spring Pleasure   62, 64
Gu Cheng Chun Se   Spring in Acient Town   153
Gu Tong Yan   Apprearance of Antiquated Brass   145
Gu Yuan Yi Feng   Relic from Classical Gardens   167
Guan Qun Fang   Fragrant Champion   204
Guan Shi Mo Yu   No 1 Jet in the World   186
Guan Yin Mian   Mother Buddha's Face   187
Gui Fei Cha Cui   Superior Imperial Concubine with Kingfisher Feather Headwear   118

H

Hai Bo   Sea Waves   154, 155
Hai Tang Hong   Chinese Crabapple Red   60
He Ding Hong   Ruby Red Crown on a Crane   91
He Yuan Hong   Red in Heyuan Garden   106
Hei Hai Jin Long   Golden Dragon in Black Sea   56
Hei Hua Kui   Black Flower Chief   85
Hong Bao Shi   Ruby   112
Hong Cai Qiu   Red Silk Ball Made of Ribbons   188
Hong Fu   Great Happiness   100
Hong He   Azaleine Lotus   64, 65
Hong Hui   Azarin Glows   123, 125
Hong Hua Lu Shuang   Frost of Freezing Dew on Scarlet Flowers   116
Hong Lian Man Tang   Carmine Lotus Dominating a Pond   62, 63
Hong Ling   Red Silk   162
Hong Ling Yan   Charming Damask Silk   70
Hong Lou Chun Hui   Spring Splendour on Magenta Mansion   119
Hong Mei Ao Xue   Proud Red Plum Blossoms in Flying Snow   161
Hong Mei Bao Chun   Plum Blossoms Announcing the Spring   170, 171
Hong Mei Fei Xue   Dancing Snow Accompanying Plum Blossoms   170, 171
Hong Mei Gui   Red Rose   102
Hong Shan Hu   Crimson Coral   108
Hong Tu   A Grand Prospect   118
Hong Xia   Puniceus Glows   65
Hong Xia Hui   Painting With Rosy Rays   191
Hong Xia Ying Ri   Prunosus Sunglows Greeting the Sun   127
Hong Xia Zheng Hui   Red Sunlight Rivaling for Fascination   110
Hong Yan Bi Rui   Pink Petals and Viridescent Pistil   108
Hong Yan He   Light Red Lotus   62
Hong Yan Yan   Brilliant Crimson   99
Hong Yu   Rubine Jade   108
Hong Yun   Madder Clouds   94, 95
Hong Yu Lou   Agate Tower   206
Hong Yun Fei Pian   Purple Pieces of Floating Clouds   69
Hong Zhu Nü   Lady in Pearl Jewellery   157
Hu Hong   Hu Hermosa   165, 166
Hua Yuan Hong   Liseran Purple in Gardens   100
Huang Cui Yu   Faint Yellow Feathers of the Kingfisher   142
Huang Hua Kui   Yellow Flower Queen   61
Huo Lian Bi Yu   Fire Refining Jasper   119

J

Ji Zhao Hong   Rooster Claw Red   160
Jia Ge Jin Zi   Spirt Gejin's Artificial Dress   205
Jia Li   Pretty Young Lady   78
Jiao Hong   Delicate Rose Hermosa   167
Jiao Nü   Ignorent Girl   62
Jiao Rong San Bian   Three Changes on Charming Appearance   198, 199
Jiao Yan   Sweet Beauty   108
Jian Rong   Broché   162, 163
Jin Gong Pao   Courtier's Robe   101
Jin Gu Chun Qing   Fine Spring in Gold Valley   141, 142
Jin Hong   Brocade of madder Purple   206, 207
Jin Lun Huang   Golden Wheel   145
Jin Pao Hong   Gown of Alizarin Red   110
Jin Si Guan Ding   Golden Silk Running through to Top   139
Jin Xing Xue Lang   Golden Stars on Snowy Waves   140
Jin Xiu Jiu Du   Splendid Jioudu   202
Jin Xiu Qiu   Brocade Ball   205
Jin Yu Jiao Zhang   Pattern Interwoven with Gold and Jade   150
Jin Yu Xi   Topaz Imperial Seal   146, 147
Jin Zhang Fu Rong   Hibiscus in Gauze Canopy   165, 166
Jing Yu   Lustre of Jade   144
Jing Yun   Beijing Aroma   86
Jiu Du Zi   Jiudu Lilac   127
Jiu Tian Lan Yue   Embracing the Moon in the Ninth Heaven   152
Jiu Zui Yang Fei   Young Yuhuan, Drunken Imperial Concubine   64
Jun Yan Hong   Smart and Bright Rose   116

K

Kun Shan Ye Guang   Night Twinkling on Mount Kunshan   138

L

Lan Bao Shi   Sapphire   117
Lan Cui Lou   Pavilion of Crimson and Sapphire   165, 166
Lan Fu Rong   Blue Lotus   120
Lan Hai Bi Bo   Clear Ripples in a Blue Sea   83, 84
Lan Hua Kui   Blue Flower Queen   194
Lan Tian Piao Xiang   Floating Fragrance in Blue Field   201
Lan Tian Yu   Lantian Jade   150
Lan Yue   Azure Moon   204
Li Hua Fen   Pear Blossom Pink   82
Li Hua Xue   Snow of Pear Blossoms   139
Li Yuan Chun   Spring in Plum Garden   128, 129
Ling Hua Xiao Cui   Green Reflections on Two-horned Flower   75
Ling Hua Zhan Lu   Limpid Dews on Two-horned Flower   119
Ling Hua Zheng Chun   Glittering Flowers Prospering in Spring   194, 195
Lu Fen   Lu Pink   150
Lu He Hong   Red Shandong Lotus   122, 124
Lü Die Wu Fen Lou   Green Butterflies Dancing around Pink Pavilion   118
Lü Mu Yin Yu   Jade Hidden in Wavy Curtain   199
Lü Xiang Qiu   Fragrant Green Ball   200
Lü Yun Fu Ri   Azure Clouds Floating around the Sun   129, 130
Luo Du Chun Yan   Spring Glamour in Luoyang Capital   114, 115
Luo Han Hong   Arhat Red   75
Luo Nü Ying   Luoyang Girl's Jade Necklace   85
Luo Nü Zhuang   Luoyang Lady with Adornments   84, 85
Luo Shen   Goddess Luo   88
Luo Yang Chun   Spring in Luoyang   152
Luo Yang Hong   Luoyang Red   106

M

Ma Ye Hong   Red of Hemp Leaves   160
Man Jiang Hong   Crimson Flooding into the River   79
Man Mian Chun   Face Radiant with Spring Happiness   135
Man Tian Xing   A Starry Sky   162
Man Yuan Chun   Spring Beauty Taking over the Garden   83, 84
Mei Gui Hong   Rosy Red   108
Mei Gui Zi   Rosy Purple   113
Ming Yuan   Town Beauty   84
Mo Jian Rong   Dark Purple Brocade   123, 125
Mo Kui   Jet Flower Queen   182, 183

Mo Lou Cang Jin   Black Ink Sprinkled with Gold   174
Mo Sa Jin   Gold Hidden in Ink-black Building   65
Mo Su   Quiet Inky Black   91
Mo Zi Rong Jin   Pale Purple and Velvety Gold   68
Mu Chun Hong   Late Spring Beauty   110,111

N

Ni Hong Huan Cai   Rainbow Radiating Brilliance   122,124

O

Ou Si Kui   Queen of Lotus-root Threads   184

P

Pan Tao Zhou   Peach Skin Crepe   98
Pan Zhong Qu Guo   Fruits Plucked from Bowl   55
Peng Sheng Zi   Prosperous Flowers in the Hand   129
Ping Hu Qiu Yue   Autumn Moon over Tranquil Lake   195,196
Ping Shi Yan   Gorgeous Duckweed Green   166,167
Po Mo Zi   Splashed Dark Purple Ink   72

Q

Qi Hua Xian Cai   Rare Flower Presenting Glory   134,135
Qi Zhu Xiang Cui   Rare Pearls Mounted in Jade   162
Qie Hua Zi   Aubergine   89
Qie Lan Dan Sha   Aubergine Blue in Red Gauze   84,85
Qin Hong   Qin Red   70
Qing Cui Lan   Indigo Blue   157
Qing Long Wo Mo Chi   Hyacinthine Dragon Lying in Ink Pool   133
Qing Long Xi Tao Hua   Dragon with Peach Blossom   188
Qing Long Zhen Bao   Purple Dragon Guarding Treasure   122
Qing Shan Guan Xue   Green Mountain Piercing the Snow   141,142
Qing Shan Wo Yun   Azure Mountains Shrouded in Clouds   122,123
Qing Xiang Bai   Delicately Fragrant White   147

R

Rou Fu Rong   Flesh-pink Hibiscus   88
Ru Hua Si Yu   An Exquisite Beauty   79
Ruan Yu Wen Xiang   Enticing Softness and Warmth   164
Ruan Zhi Lan   Pinky-blue Flower on Soft Twigs   155

S

Sa Jin Zi Yu   Alexandrite Spread with Gold   172,173
Sai Dou Zhu   Better than the Biggest Pearl   182,183
San Bian Sai Yu   Jade-like Three-coloured Flower   134,135
San Ying Shi   Three Elites   178
Shan Hu Tai   Coral Altar   158,159
Shao Nü Qun   Young Lady's Dress   108,109
Shen Zi Yu   Dark Purple Jade   74,75
Sheng Dan Lu   Blazing Furnace   204
Sheng Ge Jin   Spirit Gejin's Rival   122,124
Sheng Lan Lou   Prosperous Blue Pavilion   154
Shi Ba Hao   No.18   119
Shou An Hong   Garnet-red Light on the Desk   195,196
Shou Xing Hong   God of Longevity in Red   98
Shou Zhong Hong   Keeping Solemn Red   165,166
Shu Hua Zi   Sparse Violet Flowers   69
Shu Nü Zhuang   Fair Maiden's Costume   134,135
Si He Lian   Lotus Lookalike   69
Si Rong Hong   Velvety Purple   183
Song Bai   Song White   140,141
Song Chun   Escort of Spring   206
Song Chun Lan   Seeing off Sapphire Spring   195,196

Su Jia Hong   Su Family's Damask   89

T

Tao Hong Fei Cui   Green Flying onto Peach Blossom   119
Tao Hong Xian Mei   Peach Pink's Coquetry   204
Tao Hua Fei Xue   Peach Blossoms Fluttering Like Snow   90
Tao Hua Hong   Peach Blossom Pink   102
Tao Hua Jiao Yan   Grace and Charm of Peach Blossoms   121
Tao Hua Xi Jin   Peach Blossoms Holding Gold   123,124
Tao Hua Zheng Yan   Peach Blossoms Competing in Beauty   123,125
Tao Li Zheng Yan   Peach and Plum Trees Struggling to Bloom   110,111
Tao Yuan Chun Se   Spring in Land of Peach Blossoms   85
Tao Yuan Xian Jing   Fairyland of Peach Blossoms   94,95
Teng Hua Zi   Violet Flowers of Chinese Wistaria   182,183
Tian Ran Fu Gui   Natural Presentation of Honour and Wealth   123,125
Tian Xiang Duo Jin   Most Splendid Heavenly Fragrance   189
Tian Xiang Zhan Lu   Celestial Sweetness and Crystal Dew   154
Tian Zi Guo Se   Reigning Beauty   163
Ting Ting Yu Li   Standing Erect and Graceful   78

W

Wa Wa Mian   Baby Face   144,145
Wan Die Yun Feng   Layered Cloudy Peaks   175,176
Wan Hua Sheng   Ten Thousand Flourishing Flowers   127
Wan Shi Sheng Se   Colourful throughout the Ages   174,176
Wan Xia   Sunset Glow   82,83
Wan Xia Yu Hui   Afterglow of Nightfall   94,95
Wang Hong   Red of Royalty   196,197
Wei Hua   Flower Pavilion   170,171
Wei Zi   Lofty Purple   170
Wen Ge   Scholars' Tower   194
Wu Jin Yao Hui   Glossy Black   113
Wu Long Peng Sheng   Gragon Holding Prosperous Flowers   123,125
Wu Xia Mei Yu   Flawless Jade   143
Wu Zhou Hong   Red of Five Continents   127

X

Xi Gua Rang   Watermelon Flesh   149
Xi Shi Huan Sha   Bathing in Gauze   121
Xi Ye Zi   Narrow-leaved Purple Flower   174,175
Xia Guang Pu Zhao   Sunglow Illuminating All   190,191
Xian Chi Zheng Chun   Xianchi Prospering in Spring   64
Xian E   Fairy Maiden   134,135
Xian Gu   Celestial Woman   103
Xian Tao   Fairy Peach   203
Xiao Hu Hong   Beautiful Little Hu   164
Xiao Hu Die   Little Butterfly   78
Xiao Qing   Fine Dawn   90
Xiao Wei Zi   Little Purple Mountain   179,180
Xiang Yu   Fragrant Jade   158
Xiang Yang Da Hong   Red in Xiang Yang   178
Xiang Yang Hong   Red Facing the Sun   71
Xin Jiao Hong   Fresh and Delicate Red   118
Xing Hua Bai   Apricot Blossom White   145
Xiu Li Hong   Pretty Red   90
Xiu Rong   Cheeks Flushing Shyly   199
Xiu Tao Hua   Embroidered Peach Blossoms   203
Xiu Wai Hui Zhong   Pretty and Intelligent   164,165
Xiu Yue   The Shy Moon   77
Xue Gui   Snowy Osmanthus   141,142
Xue Li Zi Yu   Purple Jade in Snow   139
Xue Ta   Snowy Pagoda   143
Xue Ying Zhao Xia   Morning Glory Mirrored on Snow   200,201
Xu Ri   The Rising Sun   77

Y

Yan Hong Jin Bo　Golden Waves Rolling in Crimson　**85**
Yan Luo Fen He　Geese Landing on Pink Lotus　**200**
Yan Long Zi　Purple Haze at Twilight　**190**
Yan Yi Xiang Rong　Ripe Beauty and Rich Fragrance　**96**
Yan Zhi Dian Cui　Orange-red with Scattered Green　**114,115**
Yan Zhu Jian Cai　Gorgeous Pearls among Ribbons　**122,124**
Yan Zi Xian Jin　Soft Purple Touched with Gold　**91**
Yan Zi Ying Hui　Charming Purple in Reflected Light　**110**
Yang Guang　Sunbeam　**94,95**
Yang Hong Ning Hui　Red Sunshine Focusing Glory　**101**
Yao Huang　Yao's Yellow　**149**
Yi Pin Hong　Imperial Red　**106**
Yi Pin Zhu Yi　Top Rank Red Uniform　**127**
Yin Fen Jin Lin　Silver Pink and Golden Fish　**168**
Yin Gui Piao Xiang　Sweet Scented Silver Osmanthus　**140,141**
Yin Hong Lou　Silvery Red Pavilion　**164,165**
Yin Hong Qiao Dui　Silver and Red Perfectly Matched　**107**
Yin Hong Ying Yu　Silvery Red Mirroring Jade　**164,166**
Yin Lin Bi Zhu　Silver Fish Teasing Green Pearls　**175,177**
Yin Yue　Silver Moon　**190,191**
Ying Hong　Shining Red　**117**
Ying Jin Hong　Reflecting Golden Red　**106**
Ying Luo Bao Zhu　Tasselled Jades and Pearls　**206**
Ying Ri Hong　Red Welcoming The Sun　**118**
Ying Su Hong　Poppy Red Beauty　**55**
Ying Ying Cang Jiao　Hidden Beauty of Jade Lustre　**186**
Yong Chun　Praising Spring　**89**
Yue Guang　Moonlight　**154**
Yu Ban Bai　White of Jade Slab　**61**
Yu Fu Ren　Pure and Noble Lady　**190**
Yu Gu Bing Ji　Jade Bones and Ice Skin　**140**
Yu Guo Tian Qing　Clear Sky after a Shower　**160**
Yu Hou Feng Guang　Charming Sight after Rain　**105**
Yu Ji Yan Zhuang　Beauty Yuji's Gaudy Dress　**88**
Yu Lan Piao Xiang　Perfume of Yulan Magnolia　**53**
Yu Lou Chun Se　Spring Scenery In Jade Tower　**161**
Yu Lou Dian Cui　Scattered Green around Jade Pavilion　**203**
Yu Mei Ren　Jade Beauty　**137**
Yu Pan Tuo Jin　Gold on a Jade Plate　**81**
Yu Xi Ying Yue　Imperial Jade Seal Mirroring Moonlight　**151**
Yu Yi Huang　Emperor's Yellow Gown　**61**

Z

Zao Chun Hong　Early Spring Red　**72,73**
Zao Yan Hong　Early Glorious Red　**56,57**
Zhao Fen　Zhao Family's Pink　**158,159**
Zhao Jun Chu Sai　Lady Zhaojun from the Frontier Fortress　**156**
Zhao Yang Hong　Rising Sun Flaming Red　**83,84**
Zhao Zi　Zhao Family's Pruple　**166,167**
Zheng Chun　Contending in Spring　**165,166**
Zhi Hong　Cochineal Red　**119**
Zhong Sheng Hei　Traditional Black Flower　**107**
Zhong Sheng Hong　Traditional Red Flower　**206,207**
Zhong Sheng Hua　Hereditary Flower　**196,197**
Zhong Sheng Zi　Traditional Purple Flower　**175,177**
Zhou Ye Hong　Red Flower With Curly Leaflets　**169**
Zhu Hong Jue Lun　Matchless Vermilion　**172,173**
Zhu Sha Hong　Vermilion Red　**157**
Zhu Sha Kui　Hollyhock Red　**68**
Zhu Sha Lei　Red Fortress　**72**
Zhuang Yuan Hong　Top Schalor's Red Gown　**191,192**
Zi Bao Guan　Precious Purple Crown　**191**
Zi Die　Purple Butterfly　**56**
Zi Ge　Violet Tower　**122,124**
Zi Hong Cheng Yan　Beauty of the Plum Tree　**187**
Zi Hong Ling　Purplish Red Silk　**62**
Zi Hong Zheng Yan　Purplish Red Contending with Fascination　**194,195**
Zi Jin He　Purple Golden lotus　**58**
Zi Jin Pan　Purple Gold Plate　**191,193**
Zi Jing Hong　Red Flower With Purple Twigs　**134,135**
Zi Lan Kui　Purplish Blue Queen　**155**
Zi Mei You Chun　Sisters on Spring Outing　**172**
Zi Rong Jian Cai　Purple Flower with Velvet Ribbons　**195,197**
Zi Rong Qiu　Dark Purple Velvet Ball　**205**
Zi Tuo Gui　Dark Red Anemone　**132**
Zi Xia Dian Jin　Purple Glow Flickering with Gold　**184,185**
Zi Xia Ling　Purple Light on Silk　**72**
Zi Xia Xian　Fairy Maiden of Purple Light　**65**
Zi Xia Xiang Yu　Purple Rays Fringed with Jade　**65**
Zi Xia Ying Jin　Purple Sunlight Shining with Gold　**103**
Zi Yan　Purple Beauty　**91**
Zi Yan Pi Shuang　Garnet Goose Tinged with Frost　**182,183**
Zi Yang　Purple Sun　**96**
Zi Yao Tai　Purple Abode of Immortals　**184**
Zi Yu Lan　Violet Yulan Magnolia　**71**
Zi Yun　Rosy Clouds　**121**
Zi Yun Xian　Fairy Maiden of Rosy Clouds　**187**
Zui Xi Shi　Drunken Beauty Xi Shi　**167**

Editorial Committee of *Chinese Tree Peony*

Directors: Liu Dianli, Jia Xueying
Assistant Directors: Wang Lianying, Wu Yanyin, Zhang Songtao, Yang Yongchang, Duan Yunlao, Qin Kuijie
Members: Wang Changzhong, Wang Jianguo, Zhang Zuoshuang, Zhang Wantang, Song Qinghai, Li Shitian, Li Longzhang, He Xiaoyao, Wu Jingxu, Jin Zhiwei, Zhao Shujun, Zhao Xiaowu, Zhao Fudao, Jia Guilan, Han Bingzhen, Dou Yanfa, Cai Wenhou

Chief Editor: Wang Lianying
Assistant Chief Editor: Qin Kuijie
Editor Group: Wang Lianying, Liu Zheng'an, Liu Xiang, Cheng Fangyun, Zhang Shuling, Li Jiajue, Li Qingdao, Zhao Xiaozhi, Zhao Xiaoqing, Qin Kuijie, Zhang Yuexian, Lei Zengpu, Ran Dongya
Photographers: Wang Jianguo, Wang Lianying, Ran Dongya, Liu Xiang, Liu Qixian, Li Qingdao, Zhao Dixuan, Zhao Xiaozhi, Qin Kuijie
Plotters: Feng Luo, Zhou Fang, Yuan Tao
Functionaries: Ran Dongya, Zhu Shuyun, Li Zhandong, Yuan Tao

Translators: Wu Jiangmei, Chen Xinlu, Ran Dongya
English Language Consultants: Will McLewin (Phedar Nursery, Stockport, UK, SK6 3DS; 00 44 161 430 3772) and Helen Nicolson (Herunic Services, Manchester, UK, M19 2EF)

Executive editor: Chen Yingjun
Interior design: Li Qiang
Cover design: Li Zhongxin

Cooperative Units

Peony Development Administration Office of Luoyang, Henan Province; Cultural Relics and Landscape Administration Office of Luoyang, Henan Province; Peony Research Institute, Luoyang, Henan Province; Wangcheng Park, Luoyang, Henan Province; Flower and Tree Company, Luoyang, Henan Province; Scientific Research Exhibition Centre for World Precious Peony Cultivar of Luoyang, Henan Province; Peony Cultivar Propagation Base of Luoyang, Henan Province; Peony Development Company of Caozhou, Shandong Province; Zhaolou Peony Garden in Heze, Shandong Province; Peony Research Institute of Heze, Shandong; Heze Flower Garden, Shandong Province; Liji Gujin Garden, Heze, Shandong Province; Liji Peony Garden of Heze, Shandong Province; Beijing Botanical Garden, Beijing Landscape Bureau

The Authors